Manual of Water Supply Practices

M75

Elastomers for Waterworks: Pipes, Valves, and Fittings

First Edition

AF904080

American Water Works Association

Manual of Water Supply Practices—M75, First Edition

Elastomers for Waterworks: Pipes, Valves, and Fittings

If you find errors in this manual, please email books@awwa.org.
Possible errata will be posted at www. awwa.org/M75.

Managing Editor/Project Manager: Melissa Valentine
Technical Editor: Melissa Valentine
Cover Design/Technical Illustrations: Mike Labruyere
Manager, Publishing Operations: Gillian Wink
Production: Innodata
Senior Specialist - Manuals: Willadee Hitchcock
Manuals Coordinator 2: Janet Greifinger

Library of Congress Cataloging-in-Publication Data

Names: Ballun, John V. | American Water Works Association, issuing body.
Title: M75 - elastomers for waterworks : pipes, valves, and fitting / by John Ballun.
Other titles: Elastomers for waterworks : pipes, valves, and fitting
Description: First edition. | Denver, CO : American Water Works Association, 2020. | Includes bibliographical references and index. | Summary: "This manual presents the history and general application of elastomers in waterworks pipes, valves, and fittings. The principles used for evaluating, selecting, testing, and maintaining various seals will be presented"-- Provided by publisher.
Identifiers: LCCN 2020021167 | ISBN 9781647170042 (paperback)
Subjects: LCSH: Elastomers--Handbooks, manuals, etc.
Classification: LCC TS1925 .B24 2020 | DDC 628.1028/4--dc23
LC record available at https://lccn.loc.gov/2020021167

Printed in the United States of America

ISBN 978-1-64717-004-2 eISBN 978-1-61300-547-7

DOI https://doi.org/10.12999/AWWA.M75ed1

American Water Works Association

American Water Works Association
6666 West Quincy Avenue
Denver, CO 80235-3098
awwa.org

Contents

List of Figures, v

List of Tables, vii

Preface, ix

Acknowledgments, xi

Chapter 1 Introduction 1
History, 2
Purpose, 3
General Definitions, 4
References, 5

Chapter 2 Use of Elastomers 7
Linings and Coatings, 7
Static Seals, 8
Dynamic Seals, 14
Diaphragms, 17
References, 18

Chapter 3 Types and Manufacture of Elastomers 19
Types of Elastomers, 19
Selection of Elastomers, 24
Manufacture of Seals and Gaskets, 25
Manufacturing Considerations, 30
References, 31

Chapter 4 Mechanical Properties of Elastomers 33
Composition, 33
Safety, 35
Strength, 36
Resilience and Tear Resistance, 40
Temperature Resistance, 42
Test Protocols, 44
References, 45

Chapter 5 Chemical Resistance of Elastomers 47
Elastomer Chemistry, 47
Drinking Water Compliance, 49
Chlorine and Chloramine Disinfectants, 50
Elastomer Degradation Trends, 51
Permeation of Elastomers, 54
Chemical Resistance Guides, 55
Conclusion, 59
References, 59

Chapter 6 **Storage and Maintenance of Elastomers** **61**
Storage and Handling, 61
Lubrication and Maintenance, 62
Inspection, 62

Appendix A **Table of Physical/Mechanical Properties of Gaskets for Various Pipe Material Standards** **65**

Appendix B **Elastomer Test Protocols** **71**

Index, 75

List of AWWA Manuals, 79

Figures

1-1 Various elastomeric seals for the waterworks industry, 3

2-1 Check valve with rubber-lined interior and flange facing, 8
2-2 Resilient wedge gate valve with O-ring seal and rubber encapsulated wedge, 8
2-3 O-ring installation without and with pressure (P), 9
2-4 Typical waterworks flange, 10
2-5 Waterworks check valve flange connection with full-face gasket, 10
2-6 Sample full-face and ring-type gaskets, 11
2-7 PVC pipe push-on joint, 12
2-8 Ductile iron pipe push-on joint, 12
2-9 Steel pipe flanged and grooved joints, 13
2-10 Butt fusion of PE pipe, 13
2-11 Electrofusion coupling for PE pipe, 13
2-12 Mechanical joint adapter for PE pipe, 13
2-13 Mechanical joint connection, 14
2-14 Air/vacuum valve, 15
2-15 Typical dynamic seal in a dry-barrel fire hydrant, 15
2-16 Typical bolted resilient seat in AWWA butterfly valve, 15
2-17 Typical epoxy-retained resilient seats in AWWA butterfly valve, 15
2-18 Example of a knife gate valve, 16
2-19 Typical V-packing set in an AWWA butterfly valve, 17
2-20 Pinch valve, 17
2-21 Globe control valve with diaphragm, 17

3-1 Rubber ingredients loaded between two rollers for mixing, 26
3-2 Rubber lining on the interior of a butterfly valve body, 26
3-3 Typical rubber molding press with electric heating, 27
3-4 Typical transfer molding process, 28
3-5 Typical injection molding process, 28
3-6 Samples of molded rubber parts, 29
3-7 Fabrication of seals using an extrusion die, 29
3-8 Rheometer fixture, 30

4-1 Single monomer of natural rubber, 34
4-2 Vulcanization of natural rubber with sulfur, 34
4-3 Composition of NBR (nitrile), 34
4-4 Sample NSF/ANSI/CAN 61 online certification, 37
4-5 Durometer device (A scale), 38

4-6 Various durometer scales, 38
4-7 Elastomer test specimen, 39
4-8 Tension testing machine, 39
4-9 Compression set apparatus, 41
4-10 Tear resistance test specimens, 42
4-11 Typical operating temperature range of various elastomers, 43
4-12 Durometer of elastomers at reduced temperatures, 43

5-1 Elastomer degradation of a check valve disc in potable water service, 48
5-2 Degradation of a resilient gate valve wedge in service, 51

6-1 Surface degradation of an elastomeric part, 63
6-2 Portable rubber hardness tester, 63

Tables

2-1 Waterworks flange and fittings standards, 10

3-1 Example formulation for SBR compound (PHR = parts per hundred in rubber by weight), 22
3-2 ASTM D2000 definitions of material types and classes, 23
3-3 Common elastomers and generic properties, 24

4-1 Common ingredients of elastomers and their functions, 35
4-2 Required properties for a push-on pipe gasket, 37
4-3 Typical heat aging test variation for NBR (ASTM D2000), 44
4-4 Summary of standard elastomer tests, 45

5-1 Bond disassociation energy of common bonds, 49
5-2 Susceptibility of elastomers without antidegradants to chloramine attack, 52
5-3 Degradation of various generic elastomers without antidegradants in chloramine service, 53
5-4 Chemical resistance guide, 56

A-1 Physical/mechanical properties of gaskets for plastic, ductile iron, steel, and clay pipe, 66
A-2 Physical/mechanical properties of gaskets for concrete pressure pipe, 67
A-3 Physical/mechanical properties of gaskets for concrete gravity pipe, 68

This page intentionally blank.

Preface

This manual presents the history and general application of elastomers in waterworks pipes, valves, and fittings. The principles used for evaluating, selecting, testing, and maintaining various seals will be presented. It is a discussion of recommended practice, not an American Water Works Association (AWWA) standard. This document provides guidance on generally available elastomers and products and their application. Questions about specific situations or applicability to specific valves or fittings should be directed to the manufacturer or supplier.

Information contained in this manual is useful for operators, technicians, and engineers for gaining a basic understanding of the use and selection of elastomers. There are many special materials that are beyond the scope of this manual and require special consultations with suppliers and manufacturers. The elastomer properties and resistance data are generic, and actual data from the intended valve and fitting manufacturer should be obtained before making the final selection of an elastomer. This manual provides information on elastomers used in many AWWA Standards, including the following:

- AWWA C110, Ductile-Iron and Gray-Iron Fittings
- AWWA C111, Rubber-Gasket Joints for Ductile-Iron Pressure and Fittings
- AWWA C115, Flanged Ductile-Iron Pipe with Ductile-Iron or Gray-Iron Threaded Flanges
- AWWA C151, Ductile Iron Pipe, Centrifugally Cast
- AWWA C153, Ductile Iron Compact Fittings
- AWWA C200, Steel Water Pipe, 6 In. (150 mm) and Larger
- AWWA C207, Steel Pipe Flanges for Waterworks Service, Sizes 4 In. Through 144 In. (100 mm Through 3,600 mm)
- AWWA C221, Fabricated Steel Mechanical Slip-Type Expansion Joints
- AWWA C223, Fabricated Steel and Stainless-Steel Tapping Sleeves
- AWWA C227, Bolted, Split-Sleeve Couplings
- AWWA C228, Stainless-Steel Pipe Flange Joints for Water Service – Sizes 2 In. Through 72 In. (50 mm Through 1,800 mm)
- AWWA C230, Stainless-Steel Full-Encirclement Repair and Service Connection Clamps for 2-in. Through 12-in. (50-mm Through 300-mm) Pipe
- AWWA C300, Reinforced Concrete Pressure Pipe, Steel-Cylinder Type
- AWWA C301, Prestressed Concrete Pressure Pipe, Steel-Cylinder Type
- AWWA C302, Reinforced Concrete Pressure Pipe, Non-Cylinder Type
- AWWA C303, Concrete Pressure Pipe, Bar-Wrapped, Steel-Cylinder Type
- AWWA C502, Dry-Barrel Fire Hydrants
- AWWA C503, Wet-Barrel Fire Hydrants
- AWWA C504, Rubber Seated Butterfly Valves, 3 In. (75 mm) Through 72 In. (1,800 mm)
- AWWA C507, Ball Valves, 6 In. Through 60 In. (150 mm Through 1,500 mm)
- AWWA C508, Swing-Check Valves for Waterworks Service, 2-In. Through 24-In. (50-mm Through 600-mm) NPS

- AWWA C509, Resilient-Seated Gate Valves for Water Supply Service
- AWWA C510, Double Check-Valve Backflow Prevention Assembly
- AWWA C511, Reduce Pressure Principle Backflow Prevention Assembly
- AWWA C512, Air-Release, Air/Vacuum, and Combination Air Valves for Water and Wastewater Service
- AWWA C515, Reduced Wall, Resilient-Seated Gate Valves for Water Supply Service
- AWWA C516, Large-Diameter Rubber-Seated Butterfly Valves, Sizes 78 In. (2,000 mm) and Larger
- AWWA C517, Resilient-Seated Cast-Iron Eccentric Plug Valves
- AWWA C518, Double-Disc Swing Check Valves for Waterworks Service, 2-In. Through 48-In. (50-mm Through 1200-mm) NPS
- AWWA C519, High Performance Waterworks Butterfly Valves – 3 In. (75 mm) Through 60 In. (1,500 mm)
- AWWA C520, Knife Gate Valves, Sizes 2 In. (50 mm) Through 96 In. (2,400 mm)
- AWWA C606, Grooved and Shouldered Joints
- AWWA C900, Polyvinyl Chloride (PVC) Pressure Pipe and Fabricated Fittings, 4 In. Through 60 In. (100 mm Through 1,500 mm)

This manual refers to AWWA standards, which are available for purchase from the AWWA Bookstore by calling (800) 926-7337 or online at www.awwa.org/bookstore.

Manufacturers graciously provided illustrations and other documentation. AWWA does not endorse any manufacturer's products, and the names of the manufacturers have been removed from the material provided.

Metrification Note: Valve and fitting sizes are listed in their current US designation—inches. To obtain an approximate metric equivalent, use a conversion factor of 25.4 mm per inch.

Acknowledgments

The M75 Elastomers Subcommittee, which developed and reviewed this manual, had the following personnel at the time:

John Ballun, *Chair*

J.V. Ballun, Val-Matic Valve & Mfg. Corp., Elmhurst, Ill.
T. Carson, Specification Rubber Products, Inc., Alabaster, Ala.
L.R. Dunn, Bessemer, Ala.
J.A. Fancher, ACIPCO, Birmingham, Ala.
J. Grant, DeZURIK APCO HILTON, Sartell, Minn.
T.C. Harbour, Mueller Co., Decatur, Ill.
J.R. Holstrom, Val-Matic Valve & Mfg. Corp., Addison, Ill.
J.D. Larson, Saint Louis, Mo.
M. MacConnell, Burnaby, B.C.
S. Rahman, BGE Inc., Fort Worth, Tex.
E.S. Ralph, Standards Engineer Liaison, AWWA, Denver, Colo.
T.D. Rockaway, University of Louisville, Louisville, Ky.
D.B. Scott, American Flow Control, Birmingham, Ala.
J.H. Wilber, American AVK, Littleton, Colo.

The AWWA Gasket Materials for Water Works Service Committee, which reviewed and approved this manual, had the following personnel at the time of approval:

Donovan Larson, *Chair*
David Hughes, *Secretary*

General Interest Members

M.S. Alvey, Owensboro, Ky.
J.D. Larson, Saint Louis, Mo.
D.R. Purkiss, NSF International, Ann Arbor, Mich.
S. Rahman, BGE Inc., Fort Worth, Tex.
E.S. Ralph (*liaison, nonvoting*), Standards Engineer Liaison, AWWA, Denver, Colo.
T.D. Rockaway, University of Louisville, Louisville, Ky.
R.F. Williams, AECOM, Miami, Fla.

Producer Members

J.V. Ballun, Val-Matic Valve & Mfg. Corp., Elmhurst, Ill.

T. Carson (*alternate*), Specification Rubber Products, Inc., Alabaster, Ala.

S.E. Cooper (*nonvoting*), Uni-Bell PVC Pipe Association, Louisville, Ky.

L.R. Dunn (*alternate*), Besemer, Ala.

P. Gifford (*nonvoting*), Mueller Co., Chattanooga, Tenn.

T. Heartwell (*nonvoting*), VSP Technologies, Prince George, Va.

D.A. Hughes, US Pipe and Foundry, Bessemer, Ala.

K. Johnson (*nonvoting*), M & H Valve Co., Anniston, Ala.

R.L. Larkin (*nonvoting*), J & S Valve, Gardendale, Ala.

L. Mattina, Garlock Inc., Palmyra, N.Y.

D.B. Scott, American Flow Control, Birmingham, Ala.

J.H. Wilber, American AVK, Littleton, Colo.

User Members

A. Bodulow, Los Angeles Water & Power, Los Angeles, Calif.

M. Garcia (*liaison, nonvoting*), Standards Council Liaison, Denver Water, Denver, Colo.

S. Hattan, Tarrant Regional Water District, Fort Worth, Tex.

M.P. Jaspering, Missouri American Water, St. Louis, Mo.

M. MacConnell, Burnaby, B.C.

M. Spinner, Denver Water, Denver, Colo.

E. Walls, Louisville Water, Louisville, Ky.

Chapter 1

Introduction

Elastomeric linings, gaskets, diaphragms, and seals play an essential role in the operation of water distribution systems. They are used in pipes, valves, and fittings to provide tight seals for joints and valve closure members under various operating parameters including pressure, temperature, and chemical environments. There are no defining industry or AWWA material standards for specific elastomers used in the waterworks industry. Therefore, it is important to understand the chemistry, properties, and test methods associated with elastomers when selecting and specifying their use.

All materials can degrade while in service. The physical changes due to the environment, ultraviolet light exposure, wear, corrosion, chemical compatibility, fatigue, and temperature are understood for most metallic components. For elastomers, these changes are not as well cataloged and understood and can be substantially different for various and combined service conditions. Degradation and other property changes can impact expected service life and preventive maintenance frequencies. Because elastomers have different rates of degradation, the intended service life and operational compatibility are always important for proper elastomer selection.

A marked change occurred in the US waterworks industry in the 1980s when many water utilities changed from chlorine to chloramine disinfection to meet new trihalomethane (THM) standards for drinking water. Within a few months of the conversion, Ft. Lauderdale, Fla., reported rubber flappers in toilet tanks breaking down; Topeka, Kans., reported gaskets and diaphragms on backflow prevention assemblies beginning to deteriorate; and in Ingleside, Tex., the seating materials of filter backwash valves failed prematurely. These events led to several studies in the industry to better understand commonly used elastomers and their applications in the waterworks industry (Reiber 1993).

Product application and service history are important factors for elastomer selection. The selection of an elastomer for use as a dynamic seal in a safety-related valve for a nuclear power plant is much more critical than the elastomer used for a residential water heater relief valve. However, components of water systems are expected to perform as designed for many decades. Operating service experience with elastomers over long periods of time is one predictor of equipment functionality and longevity.

Many natural and synthetic elastomers are available to the designer. Some of the generic classifications of elastomers used in waterworks service include the following:

- AU, EU – Polyester and Polyether Urethane (Polyurethane)
- CR – Chloroprene (Neoprene)
- CSM – Chlorosulfonated Polyethylene (Hypalon®*)
- EPDM – Ethylene Propylene Diene Monomer
- FKM – Fluorocarbon (Viton®†)
- IIR – Isobutene Isoprene (Butyl)
- NBR – Acrylonitrile Butadiene (Nitrile, Buna-N)
- NR, IR – Polyisoprene (Natural Rubber, Synthetic Rubber)
- SBR – Styrene Butadiene
- VMQ – Polysiloxane (Silicone)

Elastomer chemistries continue to change, and many elastomer applications should be analyzed before installation. The intent of this manual is to provide a basic understanding of elastomers for the waterworks service industry.

ASTM International publishes a classification system for elastomers, which will also be explained in Chapter 3. Detailed descriptions, common uses, and temperature ratings for each of the listed elastomers will also be presented. Elastomers are enhanced with additives to improve performance and manufacturability. Unfortunately, formulations of elastomers and their additives are considered proprietary and are not published. When selecting elastomers, resistance to certain chemicals should be considered, and a chemical resistance guide is furnished in Chapter 5.

Elastomers can be mixed, molded, and extruded through various manufacturing processes to create elastomeric components or valves with integral elastomeric seals. During the manufacturing process, the properties of the elastomer can be verified with specialized quality control equipment such as a durometer and a rheometer. Finally, finished elastomer batches can be checked for many mechanical properties including hardness, tensile strength, resilience, and temperature resistance in accordance with ASTM test protocols described in Chapter 4 and Appendix B.

HISTORY

For more than 100 years, elastomeric materials have been used ubiquitously in drinking water distribution systems. Elastomeric materials have successfully been used in a variety of applications ranging from gaskets to equipment seals and have been equally suitable for a wide variety of piping gasket materials, including concrete, cast iron, ductile iron, steel, copper, and polyvinyl chloride (PVC). In general, elastomers have usually performed well in water distribution systems.

While rubber may seem like an old material, its use is actually quite recent as compared to the use, knowledge, and understanding of the basic metallic materials used in the waterworks industry. Just as there are various types of metals such as iron, carbon steel, stainless steel, lead, and bronze, there are many different elastomers like natural rubber, nitrile, styrene butadiene, ethylene propylene diene monomer, fluorocarbon, silicon, neoprene, butyl, and polyurethane. And just as there are many different alloys of the

* Hypalon® is a registered trademark of DuPont.
† Viton® is a registered trademark of The Chemours Company.

metals, there are many different chemistries and vulcanization procedures for elastomers. While the chemistries of metal alloys are well documented in published standards and are generally nonproprietary, elastomeric compounds are not. They vary widely and, in most cases, are considered proprietary.

The water industry has benefited from many of the advancements and achievements of elastomer research supported by the automotive industry. Once the water-sealing capabilities of vulcanized rubber were identified, the water industry slowly replaced lead and oakum with rubber seals and gaskets in pipe joints and other applications. While many new synthetic elastomer materials were made available, the water industry primarily focused on the "water sealing" properties of the elastomers and did not typically differentiate between materials. As such, vulcanized natural rubber was used in water system construction into the 1950s and 1960s. After that time, however, United States drinking water regulations forced a change from natural rubber to synthetic elastomer products because natural rubber appeared to promote the growth of bacteria (Leeflang 1968).

To date, elastomers and thermoplastics are revolutionizing the manufacturing industry; they appear as gaskets, coatings, membranes, hinges, structural materials, shock absorbers, and rigid bodies. They have become indispensable to the manufacturing industry and, in particular, to the portion of the industry producing valves, pipes, fittings, and the host of appurtenances required by the waterworks field (see Figure 1-1).

PURPOSE

The purpose of this manual is to explain the generic elastomer types, including their properties, characteristics, and compatibilities with waterworks systems, to assist in the selection of and application in waterworks products. Users will be able to evaluate various elastomeric materials for specific applications through a better understanding of how elastomeric parts are manufactured and tested, and a clearer understanding of their physical and mechanical properties such as strength, resilience, and hardness. Finally, the proper storage and maintenance of elastomeric parts and products are presented to help users effectively maintain inventories.

Figure 1-1 Various elastomeric seals for the waterworks industry

GENERAL DEFINITIONS

Adhesion	The state in which two surfaces are held together by interfacial forces that may consist of molecular forces, interlocking action, or both.
Age resistance	The ability of a rubber material to resist degradation of its properties caused by long-term exposure to various water environments, oxygen, heat, light, and ozone during storage or use.
Antidegradant	A compounding material used to retard degradation caused by oxidation, ozone, light, and a combination of these (ASTM D1566).
Autoclave	A pressure vessel used for the curing or vulcanization of elastomeric parts using steam under pressure.
Carbon black	A black pigment and reinforcement filler used in many rubber compounds.
Checking or crazing	A series of fine, shallow cracks appearing on a surface, resulting from aging or degradation from environmental conditions.
Chemical resistance	The ability of a material to maintain its properties while exposed to a chemical environment.
Cracking	Presence of visually noticeable deep cracks or fissures extending into the interior of a structure.
Compound	A term applied to a mixture of polymers, fillers, curing agents, and additives to produce an elastomeric material.
Compression set	The permanent deformation of a material that remains when an applied force is removed.
Copolymer	A polymer formed from two or more types of monomers, e.g., butadiene and acrylonitrile (NBR).
Degradation	The change in the physical properties of an elastomer through environmental exposure over time.
Disinfectant	Chemicals, such as chlorine and chloramines, added to drinking water to destroy microorganisms, control tastes and odors, oxidize inorganics and organics, and suppress microbiological growth in distribution systems.
Durometer	An instrument that measures a material's hardness. A material's hardness can be defined as its resistance to permanent deformation. It is also used to measure the viscous and elastic characteristics of a material when undergoing deformation. Most elastomers used in the waterworks industry are measured based on the ASTM D2240 Shore Type A and Type D scales.
Elastomer	A polymeric material with viscoelasticity (having both viscosity and elasticity) and very weak inter molecular forces, generally having low Young's modulus and high failure strain compared to other materials.
Flashing	Thin film of excess rubber formed at mold mating lines or venting gaps.
Formulation	Ingredients and their proportions in a compound; also called a recipe.
Mill	A machine used for the mechanical mixing or working of rubber consisting of two adjacent, heavy, hardened-steel, counter-rotating horizontal rolls.
Oakum	Loose fiber obtained by untwisting old rope, tarred, and used to seal gaps.
Permeation	The passage of contaminants external to the pipe, through porous, nonmetallic materials, into drinking water (USEPA 2002).
Plastic	A polymeric material that, during its manufacture or processing, can be extruded, molded, cast, drawn, or laminated into objects of all sizes and shapes by the application of heat and pressure. A thermoplastic (e.g., acrylic) can be softened and reshaped by heating and hardened by cooling. A thermoset plastic (e.g., epoxy) is irreversibly cured or hardened into a permanent shape (McTigue 2010).
Polymer	A synthetic organic compound with high molecular weight and composed of repeating chemical monomers (McTigue 2010).
Resilience	A measurement of a compound's ability to return quickly to its original shape after a temporary deformation.
Rubber	An elastomer capable of high elongation before fracture and returning quickly to its original shape after removal of the force causing deformation.

Tear strength	The force necessary to rip a material and to make the crack continue until it fails.
Tensile strength	A measurement of the stress required to rupture a standard test piece of elastomeric material.
Ultimate elongation	The percentage of distance a specimen can be strained to the point of failure or rupture.
Vulcanization	The process of treating polymers chemically to give them useful properties (such as elasticity and strength) by introducing cross-links.

REFERENCES

American Society for Testing and Materials (ASTM) International. 2015. ASTM D1566—Standard Terminology Related to Rubber. Philadelphia, Pa.: ASTM International.

Leeflang, K.W.H. 1968. Biologic Degradation of Rubber Gaskets Used for Sealing Pipe Joints. *Journal AWWA*, 60:1070–1076.

McTigue, Nancy, ed. 2010 (2nd ed.). *The Water Dictionary: A Comprehensive Reference of Water Terminology*. Denver, Colo.: AWWA.

Reiber, S. 1993. Investigating the Effects of Chloramines on Elastomer Degradation. *Journal AWWA*, August.

USEPA (United States Environmental Protection Agency). 2002. *Permeation and Leaching*. Washington DC

This page intentionally blank.

Chapter 2

Use of Elastomers

Elastomeric gaskets, linings, and seals play an essential role in the operation of water distribution system components. They are used in pipes, valves, and fittings to provide tight seals for joints and valve closure members under various operating parameters, including pressures, temperatures, and chemical environments. They are also used as diaphragms, linings, and coatings. Because there are no defining AWWA standards for specific elastomers used in the waterworks industry, it is important to understand the chemistry, properties, and test methods associated with elastomers when selecting and specifying their use.

The uses of elastomers fall into four basic functional groups as follows:

- Linings and coatings
- Static seals, such as O-rings, flange gaskets, push-on gaskets, mechanical joint gaskets, and O-ring gaskets
- Dynamic seals, such as valve seats and shaft or stem packing or O-ring seals
- Diaphragms, such as valve position controllers and flexible sheets that separate adjacent expanding and contracting chambers, as in control valves

LININGS AND COATINGS

Linings and coatings provide a barrier to protect the base material from the operating fluid or environment. Elastomeric coatings also serve to form a sealing surface typically by encapsulating a wedge, disc, or plug. The interior lining can also extend over the adjoining flange face and serve as the flange gasket. In these applications, the elastomer must be compatible with the structural properties of the base metal material and function. As a lining, the elastomer primarily provides protection from corrosion, chemical attack, wear, or impact resistance. Valves, pipes, and fittings may be lined with elastomers such as NR, IR (natural rubber) or elastomers such as CR (chloroprene) (see Figure 2-1).

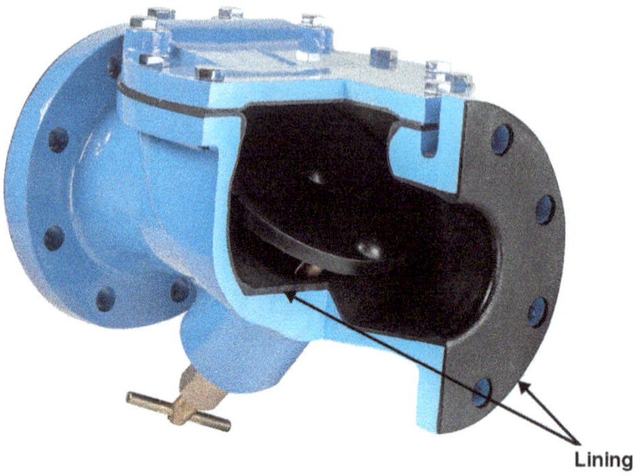

Courtesy of DeZURIK APCO HILTON.

Figure 2-1 Check valve with rubber-lined interior and flange facing

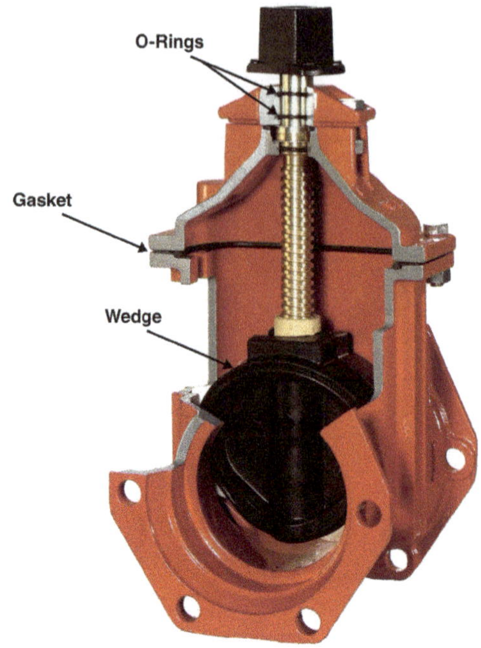

Courtesy of American Flow Control.

Figure 2-2 Resilient wedge gate valve with O-ring seal and rubber encapsulated wedge

As an encapsulating coating, elastomers can provide a sealing mechanism in various valve configurations, such as the gate valve wedge shown in Figure 2-2. Lining applications traditionally require low material strength; however, because a significant amount of surface area is exposed to the fluid and internal environment, linings and coatings are highly susceptible to fluid compatibility with the elastomer. Linings are also used for protection against abrasive media, which require a high-strength, wear-resistant elastomer.

When used as a lining or coating, the more important elastomeric properties are the long-term permeability, wear, tear resistance, and the chemical compatibility with the operating fluid.

STATIC SEALS

Static seals are installed between adjacent parts that are stationary. These include flange O-rings, flat flange gaskets, mechanical joint (MJ) gaskets, push-on gaskets, and restrained joint gaskets. In static seal applications, the elastomer is mechanically placed under compression. The seal experiences little tension, shear, or bending loads. In most water applications, thermal, pressure, and mechanical cyclic loads are generally low. Gaskets generally exhibit a low surface area exposure to the system fluid.

Static Seal – O-rings

Many valves use O-rings to seal body sections or cover plates. In Figure 2-2, the stem is sealed with O-rings. An O-ring is a molded torus or donut-shaped resilient ring typically housed in a precision gland or groove specially designed to compress the elastomeric

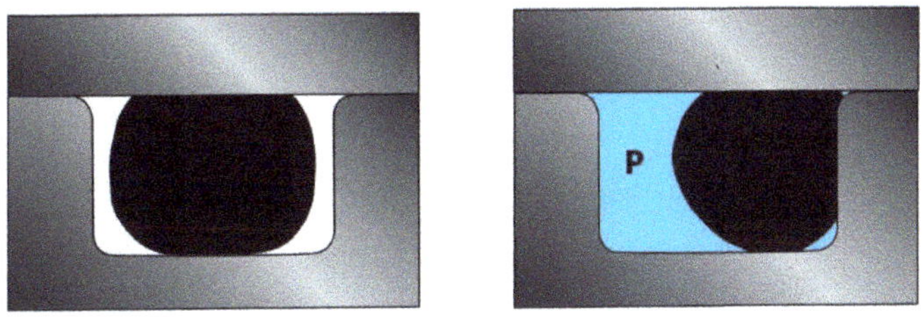

Figure 2-3 O-ring installation without and with pressure (P)

element and provide a seal. The seal is highly reliable because the groove and mating surface hold the O-ring in place, which prevents the O-ring from dislodging and leakage from occurring. The groove and mating surfaces are typically bolted together, creating metal-to-metal contact with low bolt torque to provide the necessary compression on the O-ring. As shown in Figure 2-3, the O-ring gland and groove are designed to compress the O-ring in one direction, provide volume for compression and thermal expansion, and allow the O-ring to move to one side of the groove when pressure is exerted. Because of the imposed compression, an O-ring can seal from zero to very high pressures (i.e., 0–5,000 psig or 0–34.5 MPa). While O-rings are typically static seals, they can also be used on moving parts, such as rotating shafts and oscillating cylinders, to contain the line fluid within the pressure boundary. When used in dynamic applications, the mating metal surfaces must be very smooth to prevent wear. O-rings are available in many standard circular cross sections and diameters but can also be custom-molded to suit a particular valve application. The standard sizes used by O-ring manufacturers in the United States are defined by standards such as Aerospace Standard AS-568. Typical cross-sectional diameters range from 0.040 to 0.275 in. (1.02–6.98 mm). Typical inside diameters range from 0.029 to 25.940 in. (0.74–658.9 mm). There are more than 300 standard sizes. O-rings are typically molded from high-tensile elastomers to a hardness of 70 Shore A, which is similar to that of automobile tires. O-rings are readily available in many different compounds to be compatible with hundreds of chemicals and fluids.

Static Seal – Flange Gaskets

Pipelines are often assembled using sections of pipe, fittings, and valves joined together with standardized bolted flange connections. In general, the flange connection consists of two circular flat rings of a specified thickness (T) integrally cast, welded, or threaded to the ends of the pipe with a series of bolt holes around a bolt circle (BC) with a specific diameter to compress the seal and withstand the hydrostatic and pipe loads during service (see Figure 2-4). Flanged connections are commonly used for aboveground water and wastewater service requiring a fixed restrained joint but are also used below ground in buried applications (with corrosion protection where required) and nonburied applications such as vaults. They are popular in treatment plants and industrial piping systems, in both interior and exterior applications, as well as bridge and other elevated crossings. Flanged joints allow for easier access to the piping system and associated pipeline equipment, such as valves and pumps (see Figure 2-5).

There are numerous standards for flange connections around the world with varying dimensions and pressure ratings. While there are special adapting devices available, various flange standards are not necessarily compatible either by virtue of diametrical dimensions, seal designs, bolt hole diameters, and/or bolt circle diameters. The common standards for waterworks flanges are shown in Table 2-1. Iron flanges finished in accordance with MSS SP-6 may include a flat or raised sealing surface with a serrated face (i.e., containing spiral or concentric grooves). In cases where flanges are provided with a protective coating, the

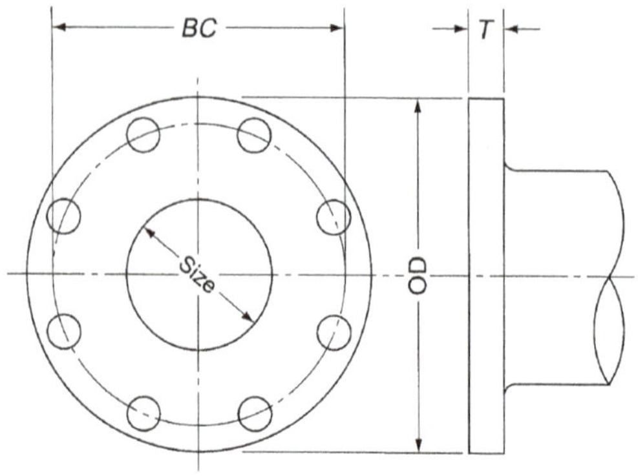

Courtesy of Val-Matic Valve & Mfg. Corp.

Figure 2-4 Typical waterworks flange

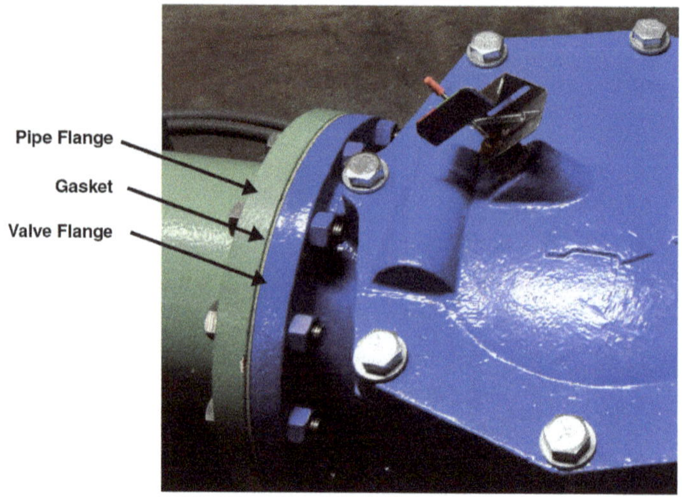

Courtesy of Val-Matic Valve & Mfg. Corp.

Figure 2-5 Waterworks check valve flange connection with full-face gasket

Table 2-1 Waterworks flange and fittings standards

Flange Standard	Materials	Classes	Size Range
ASME B16.1	Gray and ductile iron	25, 125, 250	NPS 1 to 96
ASME B16.5	Steel, stainless steel	150, 300, 400, 600, 900, 1500, 2500	NPS ½ to 24
ASME B16.42	Ductile iron	150, 300	NPS 1 to 24
ASME B16.47*	Steel, stainless steel	75A, 150A, 300A, 400A, 600A, 900A, 75B, 150B, 300B, 400B, 600B, 900B	NPS 26 to 60
AWWA C110	Gray and ductile iron	125	NPS 3 to 48
AWWA C111	Rubber gaskets	125	NPS 3 to 48
AWWA C115	Gray and ductile iron	125	NPS 3 to 64
AWWA C153	Ductile iron	125	NPS 3 to 64
AWWA C207	Steel	B (86 psi), D (175 psi), E (275 psi), F (300 psi)	NPS 4 to 144
AWWA C228	Stainless steel	SA (50 psi), SB (86 psi), SD (175 or 150 psi), SE (275 psi)	NPS 2 to 72
MSS SP-44*	Steel	150, 300, 400, 600, 900	NPS 12 to 60

Courtesy of Val-Matic Valve & Mfg. Corp.

* Caution: These standards do not match B16.1 drilling in the larger diameters. The "A" designation under "Classes" indicates MSS SP-44 compatibility; the "B" designation indicates API-605 compatibility (ASME B16.47).

surface finish requirements should only pertain to the machined surface under the coating. The machined grooves in the sealing surface help restrain the gasket from being radially forced out of the pipeline joint under the force from the internal pipeline pressure. The purpose of the gasket is to prevent leakage through the joint despite the flange's varying flatness, pressure, pipeline loads, and deflection in service.

The gasket is usually cut from sheet material to match the dimensions of the flange face or to a set of specified standard gasket dimensions. Other gasket dimensions and materials may be found in ASME B16.21 for Nonmetallic Gaskets for Pipe Flanges. Flange gaskets are available as full face, ring type, or full face with a dual-sealing element, which

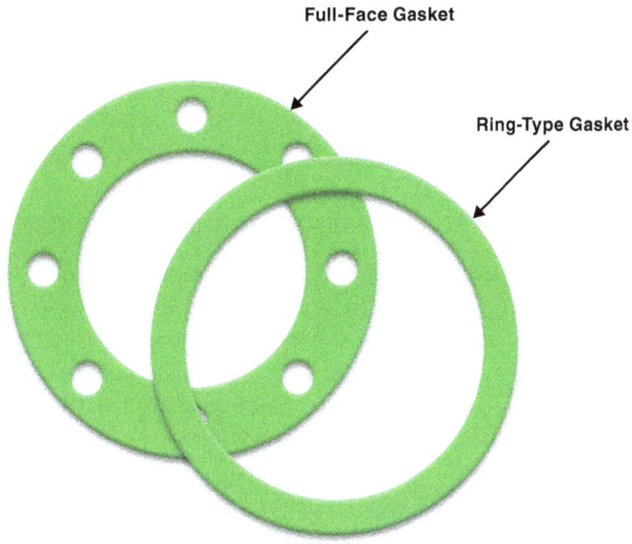

Figure 2-6 Sample full-face and ring-type gaskets

may either incorporate an integral O-ring, bulb, or seal on the inner diameter portion of the gasket (see gasket in Figure 2-2). A full-face gasket will cover the entire face of the flange and encompass the bolt holes. A ring gasket fits between the inside diameter of the flange and the inside diameter tangent edge of the bolt holes (see Figure 2-6). Ring gaskets are used for higher pressures and often require lower bolting torques given the smaller surface area of the gasket, which results in a higher gasket compression when the bolt torques are identical to full-face gaskets.

AWWA C207 describes the elastomer (rubber) gasket for steel flanged piping to be hardness (Shore A) 70–85, suitable for water service temperature of up to 200°F (93.3°C) with gasket seating stress of 200 psi (1,379 kPa) minimum. The specified hardness is similar to that of a shoe heel (Figure 4-6) and assures that the gasket maintains its dimensional shape when compressed by the flange bolts. If assembly seating stress recommendations are unavailable, elastomeric gaskets are rated for a minimum of 175 psi (1,206 kPa) internal working pressure by the manufacturer. Compressed fiber gaskets are a blend of synthetic fibers, fillers, and elastomeric binders suitable for potable water service. Gaskets that are polytetrafluoroethylene (PTFE) based are also acceptable. Gasket seating stress should be no greater than 4,800 psi (33 MPa). The specified seating stress provides the necessary strength to prevent the blowout of the gasket or damaging plastic deformation during pipeline operation.

For sealing harsh chemicals or high pressures or temperatures, the gasket material is usually a fiber gasket with a rubber binder material. Depending on the fiber and rubber binder, fiber gaskets are rated to pressures over 500 psi (3,450 kPa) and temperatures over 300°F (149°C). Some types are also meant for electrical isolation purposes, such as between differing pipe materials for corrosion prevention or cathodic protection. The gasket manufacturer should be consulted for specific temperature and chemical exposure limits.

ASME PCC-1 Guidelines for Pressure Boundary Bolted Flange Joint Assembly provides recommendations for the use and installation of ring-type gaskets for general pressure service, including recommended flange bolt tightening patterns and target torque values. The leak tightness of a bolted flanged joint assembly is related to many factors in addition to the selection of the gasket material. AWWA provides an online, interactive tool for members to evaluate steel flange connections based on AWWA C207 requirements, AWWA M11 guidelines, various gaskets materials, and bolt torques (AWWA Flange Torque Calculator).

When used as a static seal, the important elastomeric properties of the gasket are the compressive and tensile strength, compression set, permeability, and hardness. Depending on application, temperature limitations may also be important.

Static Seal – Push-On Gaskets and Pipe Connections

Pipelines are also assembled with various elastomeric seal designs depending on the pipe material (e.g., plastic, concrete, and iron) and type of joint (e.g., push-on, flanged, and mechanical joint). The performance requirements for the elastomeric gaskets are contained in either the pipe standard or a standard specific to the type of joint. A summary of the physical/mechanical properties of gaskets for the various pipe material standards are presented in Appendix A.

For polyvinyl chloride (PVC) pipe joints, a push-on gasket meeting ASTM F477 is provided (ASTM 2014). As shown in Figure 2-7, sections of PVC pipe are joined using a push-on joint where a continuous elastomeric ring gasket is compressed into an annular space formed by the pipe bell and the spigot end of the pipe. The dimensions of the socket and gasket are in accordance with the manufacturer's standard design dimensions.

Similarly, ductile iron pipe can also be furnished with bell and spigot ends sealed with a push-on gasket meeting ANSI/AWWA C111 as shown in Figure 2-8.

Steel pipe sections can be joined by welding, flanges, or grooved/shouldered joints (see Figure 2-9).

Polyethylene (PE) pipe is joined by butt fusion, electrofusion couplings, or with mechanical joining alternatives as shown in Figures 2-10, 2-11, and 2-12.

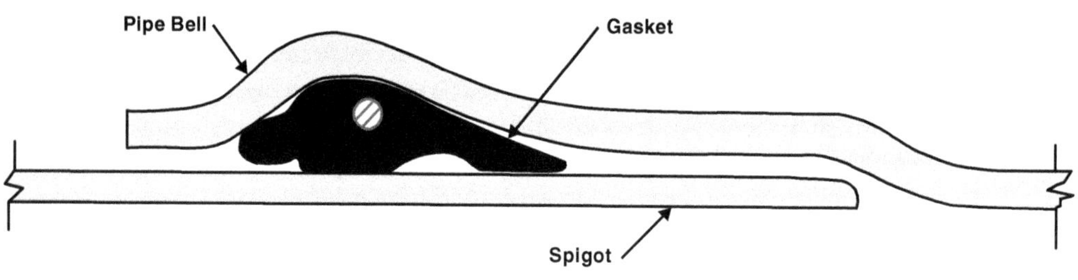

Figure 2-7 PVC pipe push-on joint

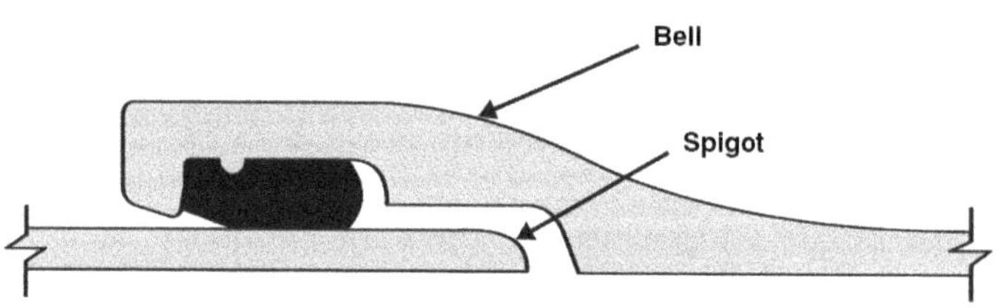

Courtesy of Ductile Iron Pipe Research Association.

Figure 2-8 Ductile iron pipe push-on joint

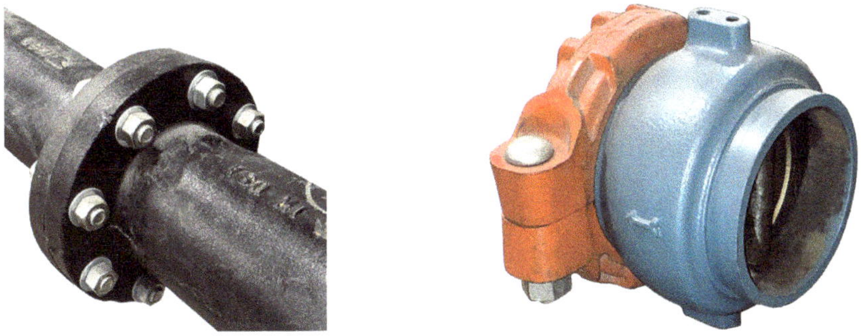

Figure 2-9 Steel pipe flanged and grooved joints

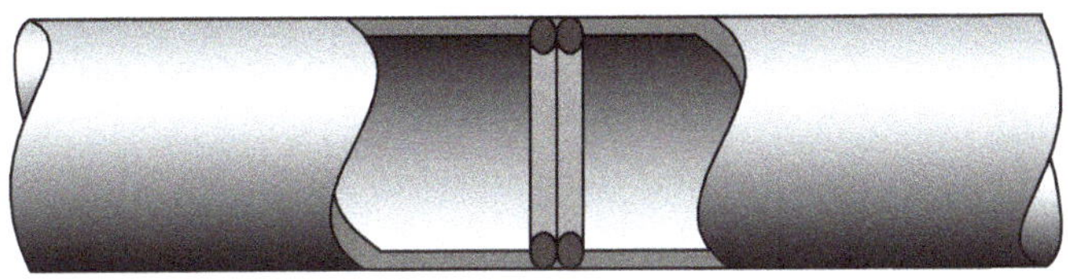

Figure 2-10 Butt fusion of PE pipe

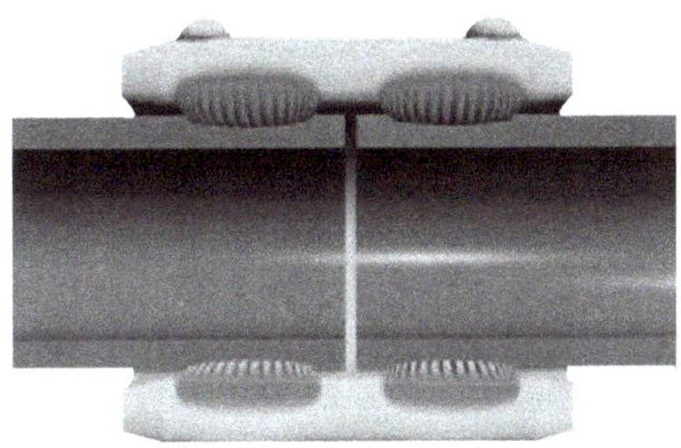

Figure 2-11 Electrofusion coupling for PE pipe

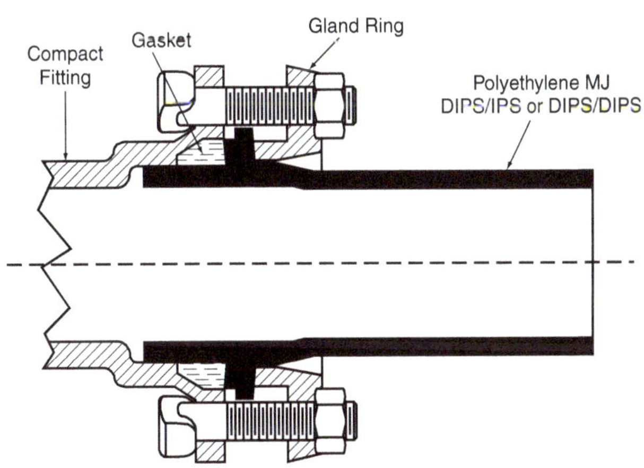

DIPS—ductile iron pipe size, IPS—iron pipe size, MJ—mechanical joint

Figure 2-12 Mechanical joint adapter for PE pipe

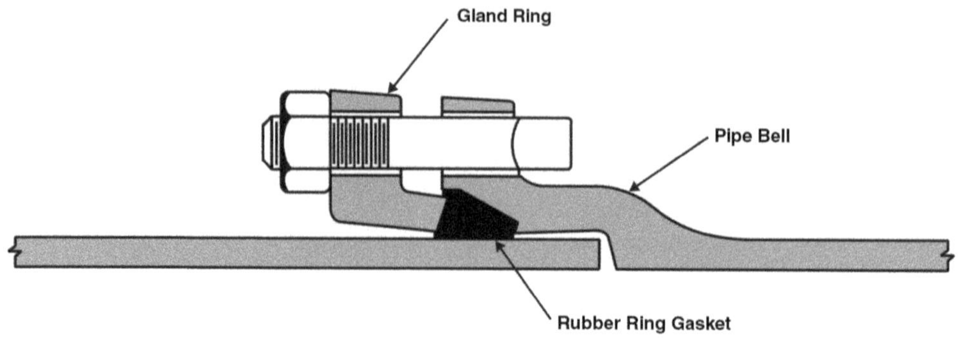

Courtesy of Ductile Iron Pipe Research Association.

Figure 2-13 Mechanical joint connection

Static Seal – Mechanical Joint Gaskets

Another common gasket joint in the waterworks industry is the mechanical joint connection where the plain pipe end is inserted into a special hub with gasket and gland ring as shown in Figure 2-13. In this joint, a gland ring compresses a specially shaped rubber ring gasket between the outside diameter of the pipe and the inside diameter of the pipe bell.

Similar to the flange joint, the important elastomeric properties of the gasket are the compressive and tensile strength, compression set, permeability, and hardness. Chemical compatibility with the operating fluid is less important given the small surface area exposure to liquid versus material volume. Depending on application, temperature and other limits specific to the gasket material type may also be important.

Static Seal – Valve Face Seal

Many types of valves rely on a flat elastomeric seal that mates to a metal surface. These seals are common in check valves and air valves, for example. Figure 2-14 illustrates an air/vacuum valve, which can be mounted on the top of water or wastewater lines. The round stainless steel float presses against a flat resilient seal to prevent water from leaking out of the valve. Similar to other static seals, the important elastomeric properties of static valve seals are compressive strength, compression set, permeability, and hardness.

DYNAMIC SEALS

Dynamic seals perform a seal function between parts that move relative to each other during operation. In this application, the mechanical loads are associated primarily with valve closure seal mechanisms. These vary widely by valve type and design, and with the type of shaft or stem packing seal. These seals must withstand the stresses under seating, unseating, and operating loads. These loads usually include compression, tension, pressure, bending, and cyclic loads. In water applications, thermal and thermal cycling stresses are generally low. Thus, surface area exposure becomes a significant factor. Finally, any degradation caused by chemical exposure can also be detrimental to the valve's ability to seal.

The important elastomeric properties of dynamic seals can vary widely given the design of the valve seal or gland but generally include most of the mechanical properties previously noted for static seals.

Valves often rely on resilient seals for tight shutoff. Figure 2-15 illustrates a common seal used in dry-barrel fire hydrants. Figures 2-16 and 2-17 illustrate typical butterfly

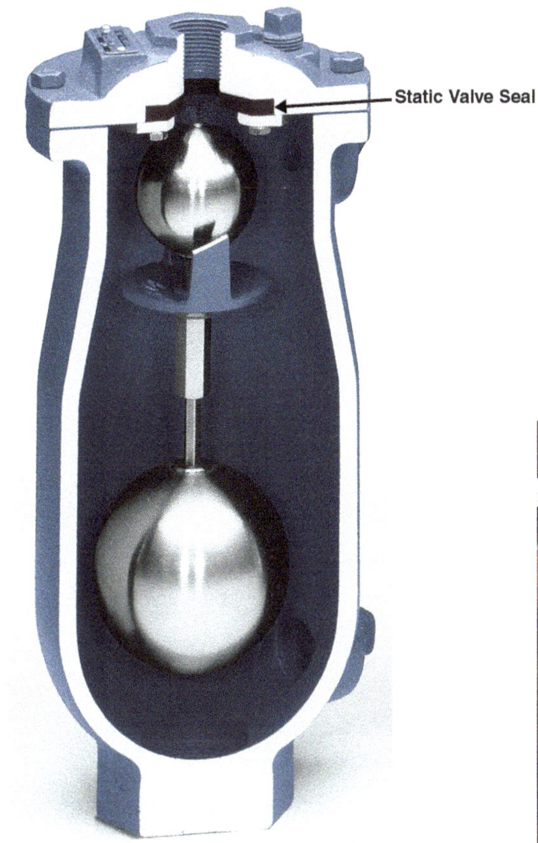

Courtesy of Val-Matic Valve & Mfg. Corp.

Figure 2-14 Air/vacuum valve

Courtesy of American Flow Control.

Figure 2-15 Typical dynamic seal in a dry-barrel fire hydrant

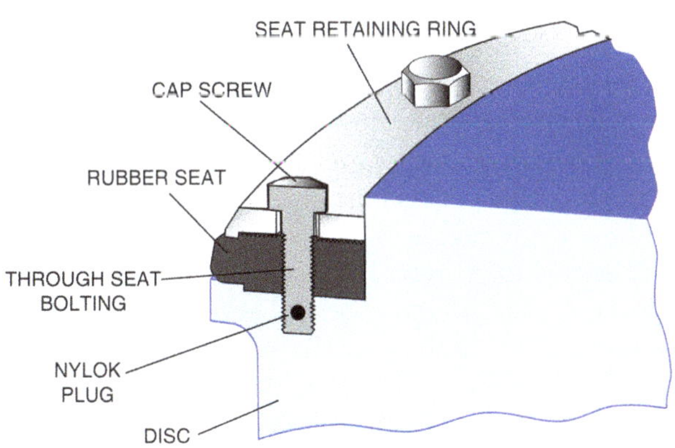

Courtesy of Val-Matic Valve & Mfg. Corp.

Figure 2-16 Typical bolted resilient seat in AWWA butterfly valve

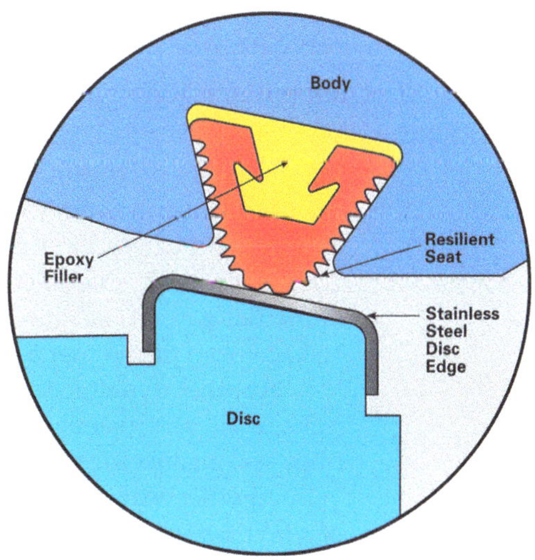

Courtesy of DeZURIK APCO HILTON.

Figure 2-17 Typical epoxy-retained resilient seat in AWWA butterfly valve

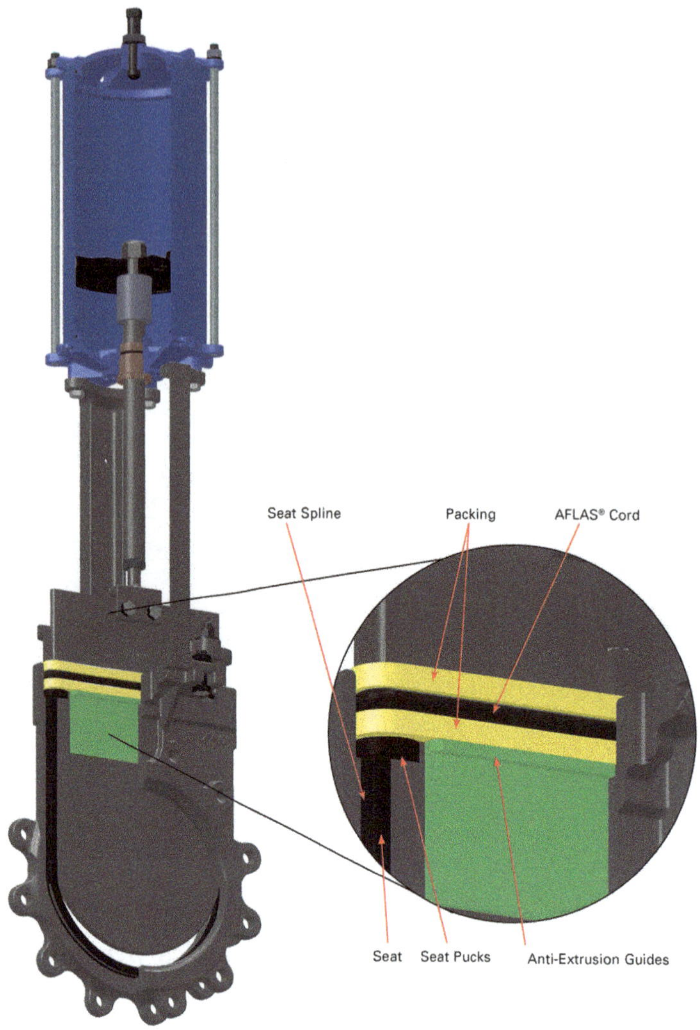

Courtesy of DeZURIK APCO HILTON.

Figure 2-18 Example of a knife gate valve

valve resilient seats mounted to the disc edge or in the valve body, and mate to stainless steel surfaces when the valves are fully closed. A butterfly valve rubber seat is subject to compression when closed and subject to shear forces while closing and opening.

Some knife gate valves have resilient seats that seal against the face or perimeter of the flat gate. A packing assembly is used to create a seal around the gate, preventing leakage from the bonnet of the valve as shown in Figure 2-18.

Another dynamic seal is valve packing, which is typically used as a barrier against leakage via protrusions in a valve body. As shown in Figure 2-19, the V-shaped packing rings seal against the shaft bore in the body and the rotating surface of the valve shaft.

Another example of a dynamic seal is the interior seal in a pinch valve, which consists of a fabric-reinforced tubular elastomeric sleeve that is compressed by a pinch bar to shut off the flow (see Figure 2-20). These valves are ideally suited for slurry flow or fluids containing solid media as there are no cavities or obstructions in the flow area for solids to collect and obstruct closure of the valve. The rubber sleeve must have high tensile strength, resilience, tear resistance, and chemical compatibility with the process fluid.

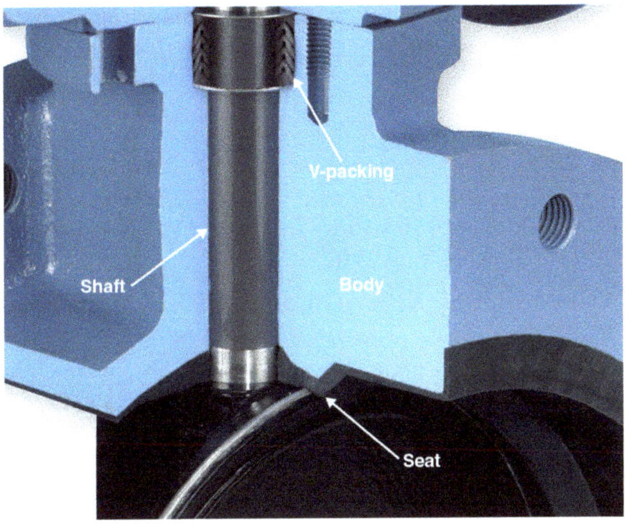

Courtesy of DeZURIK APCO HILTON.

Figure 2-19 Typical V-packing set in an AWWA butterfly valve

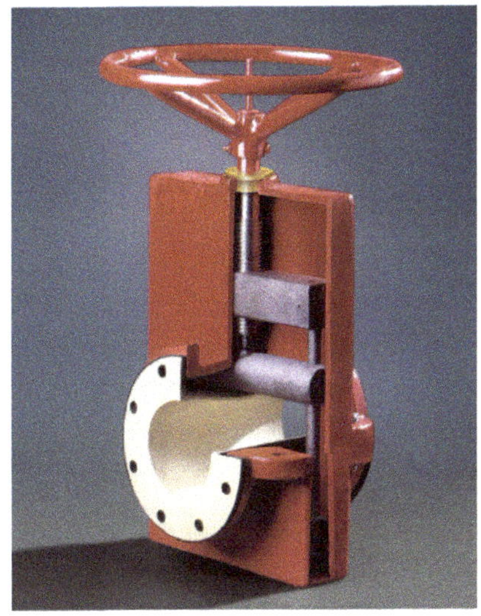

Courtesy of Red Valve.

Figure 2-20 Pinch valve

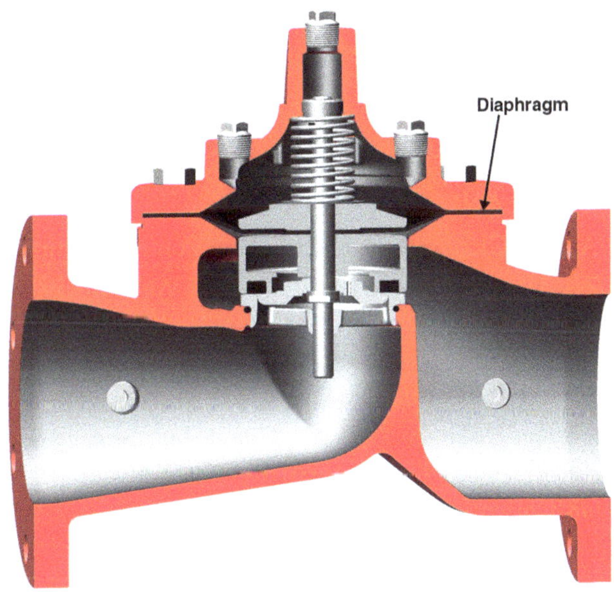

Courtesy of Cla-Val Company.

Figure 2-21 Globe control valve with diaphragm

DIAPHRAGMS

Diaphragms in valves are used to create a boundary between differing pressure cavities to assist with operation (see Figure 2-21). In a control valve application, the mechanical loads are associated primarily with valve actuation and seal operations and can vary widely by valve type and design. These diaphragms must withstand the stresses under pressure boundary loads, including compression, tension, pressure, bending, and cyclic. In water applications, thermal cycling loads are generally low.

Similar to dynamic seals, the important elastomeric properties of diaphragms vary widely by design but generally include most of the mechanical properties previously noted for static seals.

REFERENCES

American Society of Mechanical Engineers (ASME). 2015. ASME B16.1. Gray Iron Pipe Flanges and Flanged Fittings: Classes 25, 125, and 250. New York, N.Y.: ASME.

ASME. 2017. ASME B16.5. Pipe Flanges and Flange Fittings: NPS ½ Through NPS 24 Metric/Inch Standard. New York, N.Y.: ASME.

ASME. 2016. ASME B16.21. Nonmetallic Flat Gaskets for Pipe Flanges. New York, N.Y.: ASME.

ASME. 2016. ASME B16.42. Ductile Iron Pipe Flanges and Flanged Fittings: Classes 150 and 300. New York, N.Y.: ASME.

ASME. 2017. ASME B16.47. Large Diameter Steel Flanges: NPS 26 Through NPS 60 Metric/Inch Standard. New York, N.Y.: ASME.

ASME. 2013. ASME PCC-1. Guidelines for Pressure Boundary Bolted Flange Joint Assembly. New York, N.Y.: ASME.

American Society for Testing and Materials (ASTM) International. 2014. ASTM F477. Standard Specification for Elastomeric Seals (Gaskets) for Joining Plastic Pipe. Philadelphia, Pa.: ASTM International.

American Water Works Association (AWWA). 2012. ANSI/AWWA C110/A21.10. Ductile-Iron and Gray-Iron Fittings. Denver, Colo.: AWWA.

AWWA. 2012. ANSI/AWWA C111/A21.11. Rubber-Gasket Joints for Ductile-Iron Pressure Pipe and Fittings. Denver, Colo.: AWWA.

AWWA. 2011. AWWA C115. Flanged Ductile-Iron Pipe with Ductile-Iron or Gray-Iron Threaded Flanges. Denver, Colo.: AWWA.

AWWA. 2011. ANSI/AWWA C153/A21.53. Ductile-Iron Compact Fittings. Denver, Colo.: AWWA.

AWWA. 2017. ANSI/AWWA C207. Steel Pipe Flanges for Waterworks Service—Sizes 4 In. Through 144 In. Denver, Colo.: AWWA.

AWWA. ANSI/AWWA C228. Stainless-Steel Pipe Flanges for Water Service – Sizes 2 In. Through 72 In. (50 mm Through 1,800 mm). Denver, Colo.: AWWA.

AWWA. 2018. *Flange Torque Calculator*, AWWA Flange Bolt Stress and Torque Analysis, Rev. 1.0. ascelibrary.org/doi/10.1061/9780784481646.013.

Manufacturers Standardization Society (MSS). 2017. MSS SP-6. Standard Finishes for Contact Faces of Pipe Flanges and Connecting-End Flanges of Valves and Fittings. Vienna, Va.: MSS.

MSS. 2016. MSS SP-44. Steel Pipeline Flanges. Vienna, Va.: MSS.

SAE International. 2020. Aerospace Size Standard for O-Rings.

Society of Automotive Engineers (SAE) International. 2014. SAE AS568D. Aerospace Size Standard for O-Rings. Warrendale, Pa.: SAE International.

Chapter 3

Types and Manufacture of Elastomers

Elastomeric gaskets and seals play an essential role in the operation of water distribution system equipment. Without gaskets and seals, piping systems and equipment would not maintain their integrity and contain fluid within their internal pressure boundary. Rubber can be molded, extruded, or machined into almost any shape imaginable to match the application or piece of equipment. Given the wide array of possible elastomer properties and chemical resistances available, there are a multitude of base elastomeric compounds in use. Moreover, the basic elastomeric compounds are enhanced with additives to provide improved mechanical and environmental resistance to match the intended application. An understanding of the basic elastomers and their manufacture is important for their proper selection and application.

TYPES OF ELASTOMERS

There are more than 25 base polymers used in US manufacturing, and from these, thousands of elastomeric formulations are possible. Dozens of base polymers are commercially available with properties that vary in terms of elasticity, hardness, temperature resistance, and environmental resistance. Engineered compounds are created by manufacturers for special applications by combining the basic polymers with fillers, accelerators, activators, agents, and special additives. The number of possible engineered compounds is endless, and these resultant compounds are considered proprietary. Their ingredients are not generally published.

The elastomer types commonly used in the water industry are much more limited, generally breaking down into categories reflecting eight base polymers (Rockaway et al. 2007). Polyester urethane (AU) and chlorosulfonated polyethylene (CSM) are also commonly used and described (Harper 1992) below:

AU, EU – Polyester and polyether urethane (Polyurethane)

AU, EU is compounded with a variety of properties to act as a resilient elastomer for gaskets and padding or a hard and durable material for rollers, wheels, and molds. It does not support fungus growth and is generally resistant to microbiological attack. It is limited in temperature to typically 175°F (80°C) but has excellent wear resistance, high tensile strength, and high elasticity in comparison with any other elastomers. AU, EU is compatible with ozone, water, mineral, and vegetable oils, petroleum hydrocarbons, and greases but should not be used for chlorinated solvents, hot water, or steam. It is softened and swollen by alcohols. Within the water industry, AU, EU is generally not resistant to water containing residual disinfectant chemicals such as chlorine and chloramines (Pruett 1994).

CR – Chloroprene rubber (Neoprene)

CR is one of the best all-purpose elastomers for unique conditions where resistance to sunlight, oxidation, heat, and many petroleum derivatives is of prime importance. It has a high degree of flame resistance because of its chlorine content, and it provides excellent resistance to many oils, chemicals, water acids, and alkalis. It is generally attacked by petroleum hydrocarbons and chlorinated solvents (Pruett 1994). CR is used in a wide variety of environments, such as in wetsuits, electrical insulation, and car fan belts. Its chemical inertness makes it well suited for industrial applications such as gaskets, hoses, and corrosion-resistant coatings. Within the water industry, CR is generally not resistant to water containing residual disinfectant chemicals such as chlorine and chloramines (Rockaway et al. 2007).

CSM – Chlorosulfonated polyethylene (Hypalon®)

CSM has good resistance to oxygen, ozone, and ultraviolet (UV) light. It is also oil and grease resistant and has low gas permeability. It is generally attacked by petroleum hydrocarbons and chlorinated solvents (Pruett 1994). Because of its excellent UV stability, CSM is useful as roof sheeting and for reservoir liners, as well as wire and cable applications, coated fabrics, and hoses. CSM is one of the most common materials used to make inflatable boats and is also used as a single-ply roofing membrane, and as a surface coat material on radomes due to its radar-transparent quality. Within the water industry, CSM is generally not resistant to water containing residual disinfectant chemicals such as chlorine and chloramines (Pruett 1994).

EPDM – Ethylene propylene diene monomer

EPDM is a relatively new elastomer product formulated by ExxonMobil Chemical in the 1960s. EPDM rubber is ideal for outdoor applications, such as a roofing material, because of its excellent resistance to ozone, oxidants, and severe weather conditions. It is generally attacked by petroleum hydrocarbons and chlorinated solvents (Pruett 1994). Its primary limitation is its susceptibility to hydrocarbons. Within the water industry, EPDM is resistant to water containing residual disinfectant chemicals such as chlorine and chloramines. When cured with peroxide, EPDM exhibits an improved chemical and thermal resistance compared to that which is cured with sulfur. However, sulfur-cured EPDM exhibits better mechanical and physical properties, including higher tensile strength and a greater resistance to tearing. When selecting rubber for an application, consideration should be given to all variables that can impact elastomer performance. EPDM is one of several elastomers widely used for O-rings, flat gaskets, valve and hydrant seats, and pond liners.

FKM – Fluorocarbon (Viton®)

Because of its excellent chemical resistance and outstanding upper temperature limit of 390°F (200°C), FKM has become one of the most widely used seal materials in the world. The material is resistant to the majority of greases, fuels, oils, petroleum hydrocarbons, and chlorinated solvents. Within the water industry, FKM is generally resistant to water containing residual disinfectant chemicals such as chlorine and chloramines. The high cost of FKM is usually justified by its all-around performance. Fluorocarbon is used for O-rings, gaskets, seals, and peristaltic pump tubes.

IIR – Isobutene isoprene (Butyl)

IIR has been available to the industry for more than 60 years. The primary attributes of this material are excellent impermeability/air retention and good bending properties. It displays outstanding resistance to many chemicals, including oxygenated solvents and ozone, and it is also resistant to water, steam, sunlight, and heat. It is generally attacked by petroleum hydrocarbons and chlorinated solvents (Pruett 1994). Within the water industry, IIR is resistant to water containing residual disinfectant chemicals such as chlorine and chloramines (Reiber 1993). IIR is primarily used for diaphragms, construction sealants, and hoses.

NBR – Acrylonitrile butadiene (Nitrile, Buna-N)

NBR is an oil-resistant elastomer that was formulated shortly after World War II. In addition to its resistance to petroleum hydrocarbons and chlorinated solvents, NBR is highly resistant to mineral oils, vegetable oils, and many acids. For this reason, it has wide application in machinery and instrumentation. It also has good elongation properties as well as adequate resilience, tensile, and compression set. While typically more expensive than SBR, municipal utilities frequently use NBR where a combination of oil and water residues is anticipated, such as in sewage. Within the water industry, NBR is generally not resistant to water containing residual disinfectant chemicals such as chlorine and chloramines (Rockaway et al. 2007).

Typical application for NBR includes seals, O-rings, gaskets, diaphragms, tank linings, boots, and valve seats (Reiber 1993).

NR, IR – Polyisoprene (Natural rubber, synthetic rubber)

NR, natural rubber, is compounded from gum stock harvested from trees, much like maple syrup. It was the first elastomeric product adapted for use in the water distribution industry. It was relatively easy to form and was effective in creating watertight products. NR has been generally replaced with IR, synthetic polyisoprene, in current water distribution systems. IR is a synthetic elastomer developed for use as a substitute for NR. Like NR, IR is susceptible to ozone and disinfectant-induced degradation as well as to biological attack. It is attacked by petroleum hydrocarbons and chlorinated solvents (Pruett 1994). Within the water industry, NR and IR are not resistant to water containing residual disinfectant chemicals such as chlorine and chloramines (Reiber 1993). IR would otherwise be a good gasket material due to excellent resilience, tear strength, and wear resistance. IR compounds are commonly used in static concrete pressure pipe gaskets where gasket exposure to disinfectants is minimal. Other suitable applications include heavy-duty tires, vibration dampers, gaskets, seals, and rollers.

SBR – Styrene butadiene

SBR was developed during World War II as a synthetic alternative to replace NR. SBR is a low-cost material that has similar chemical and physical properties as NR, including poor ozone resistance, with generally better abrasion resistance. It can be compounded to provide very fine abrasion, wear, and tensile qualities. However, it is susceptible to oxidant attack and has poor resistance to the majority of organic solvents. It is generally attacked by petroleum hydrocarbons and chlorinated solvents (Pruett 1994) but has been studied and accepted for use as pipe gaskets in contaminated groundwater conditions (see Chapter 5). Within the water industry, SBR is not resistant to water containing residual disinfectant chemicals such as chlorine and chloramines (Reiber 1993) but is commonly used for pipe gaskets where there is a low surface area–mass ratio and swelling does not affect performance (Rockaway et al. 2007). In the water industry, SBR is commonly used in pipe gaskets and other water applications where only a portion of the part is exposed to potable water.

VMQ – Polysiloxane (Silicone)

VMQ has many unique properties that most organic elastomers are unable to copy. One excellent example is the wide operating temperature range, from a very low temperature of –58°F (–50°C) to a very high temperature of 480°F (250°C). VMQ is also oil resistant and resists abrasion well. It is generally attacked by petroleum hydrocarbons and chlorinated solvents (Pruett 1994). Within the water industry, VMQ is generally resistant to water containing residual disinfectant chemicals such as chlorine and chloramines (Reiber 1993). Its high temperature resistance makes VMQ an excellent choice for many automotive, industrial, and high-temperature applications, such as a kitchen spatula. However, due to its expense, VMQ has seen limited application in the water industry.

Most elastomer manufacturers do not share their research regarding "best design" compounds with the public. Valve and pipe manufacturers balance desired elastomer properties and characteristics with those elastomers that are available and affordable. Therefore, a user of such products has limited industry reference material to research and form an educated choice. An example SBR compound is given in Table 3-1. This formulation was one of the seven used in an AwwaRF-sponsored research project (Rockaway et al. 2007) to provide some insights into elastomer degradation by analyzing commercially available elastomers in the industry. The study reported on the advantages and disadvantages of each elastomer. To mimic the effects of long-term use, higher temperatures and higher concentrations of disinfectants were applied to the samples.

The ingredient model numbers for this specific mixture in Table 3-2 are specific to a particular chemical company. There are hundreds of other additives that could be included to improve the extrusion or molding process, or enhance the compound's resistance to specific environmental exposures, such as ozone, UV light, aging, and temperature. As a result, some seal and equipment manufacturers (e.g., Parker Hannifin Corp. [2018] and Blair Rubber Company [2018]) list their formulations or compounds by model number and publish general mechanical and environmental properties for the compounds.

There is currently one reference standard that provides detailed elastomer information and classification. American Society for Testing and Materials (ASTM) International publishes a classification of the properties of elastomeric compounds in ASTM D2000. ASTM D2000 is a classification system for the properties of vulcanized rubber compounds. The system is used to arrange rubber products based on characteristics. The rubber products are arranged by type based on resistance to aging and class based on resistance to oil swell. Basic requirements are established for each type and class along with additional requirements to reflect the intended purpose of the material.

Table 3-1 Example formulation for SBR compound

Ingredient	PHR	Function
SBR 1502	100	Styrene butadiene base polymer
SRF N767	90	Carbon black filler
Sun 4240	10	Extending oil
Sulfur	1	Primary vulcanizing agent
Zinc oxide	5	Curing system activator
Stearic acid	1	Curing system additive
Santocure	1	Accelerant

PHR = parts per hundred in rubber by weight
Adapted with permission from Water Research Foundation.

Table 3-2 ASTM D2000 definitions of material types and classes

Material Type	Test Temperature (°C)	Material Class	Maximum Swell (%)
A	70	A	N/A
B	100	B	140
C	125	C	120
D	150	D	100
E	175	E	80
F	200	F	60
G	225	G	40
H	250	H	30
J	275	J	20
K	300	K	10

Adapted from ASTM International.

The purpose of ASTM D2000 is to provide general industry guidance in the selection of the most suited rubber compound for a particular application. The development of a "line call-out" is used to specify the properties of the material so that properties are easily understood.

Below is an example of a typical line call-out for an NBR gasket compound:

M-4-B-G-7-14-A13-B13-C12-EA14

where:

M = SI unit designation
4 = Suffix grade of material
B = Material type based on heat resistance (B = 100°C (212°F))
G = Material class based on oil resistance (G < 40% swell)
7 = × 10 is Type A durometer hardness ±5
14 = Tensile strength (MPa) (2,031 psi)
A13 = Heat resistance test at 100°C (212°F)
B13 = Compression set test at 100°C (212°F)
C12 = Ozone resistance test
EA14 = Water resistance test

By referencing the ASTM D2000 specification, an engineer is able to see the test parameters and specifications for a particular elastomer type and class of rubber compound. The material type is defined by the test temperature used for a 70-hour aging test and meeting changes in tensile strength of ±30%, Shore A hardness of –50% max., and Shore A hardness of ±15 points. The material class is defined as the maximum volume swell of the material in oil after 70 hours at the temperature associated with the material's type or 150°C max. (see Table 3-2). Any special requirements that deviate from the standard references are allowed to be expressed with a "Z" designation. The "Z" designation must be clarified as to the testing and results desired, for example:

M-4-B-G-7-14-A13-B13-C12-EA14-Z1-Z2

where:

Z1 = Hardness of 70 ±5
Z2 = Compression set 20% max.

The use of the ASTM D2000 system allows for a common language to be spoken across different disciplines and industries.

SELECTION OF ELASTOMERS

Several criteria should be considered in the selection of an elastomer. In addition to cost, the function of the elastomeric part plays an essential role along with the environmental conditions. As illustrated in Chapter 2, elastomeric parts can be linings, seals, gaskets, packing, and diaphragms. For example, elastomeric valve seals and packing require an elastomer that is wear resistant. Conversely, gaskets require compressive strength and low permeability. Finally, valve diaphragms require tensile strength, low permeability, and high strain ability.

After the list of elastomers are narrowed down, based on function, the environmental conditions should be considered. Each elastomer will have a temperature rating at which point certain mechanical properties of the elastomer will be degraded (see Table 3-3). The chemical environment is another critical element of the selection process. The concentration should be known, along with the length of exposure and the surface area of the elastomeric part. Certain pipe gaskets may degrade with exposure to groundwater contaminants, but because of their design, only a small surface of the gasket may be directly affected. It is important to understand the failure mode of the elastomeric part. A butterfly valve seat

Table 3-3 Common elastomers and generic properties

Generic Elastomer	Common Environments*	Normal Temperature Rating	ASTM D2000 Designation Type, Class
AU, EU: Polyester and polyether urethane (Polyurethane)	Water, ozone, hydrocarbons, and greases. Not for use with chlorinated solvents, alcohols, hot water, or steam.	175°F (80°C)	BG
CR: Chloroprene (Neoprene)	Water, saltwater, wastewater, greasy waste, and ultraviolet (UV) light. Not for use with petroleum hydrocarbons and chlorinated solvents.	240°F (110°C)	BC, BE
CSM: Chlorosulfonated polyethylene (Hypalon®)	Water, oil, grease, ozone and outdoor service, and UV light. Not for use with hydrocarbons and chlorinated solvents.	255°F (125°C)	CE
EPDM: Ethylene propylene diene monomer	Water, saltwater, wastewater, elevated temperatures, outdoor service, UV light, dilute acids, dilute alkalis, ketones, acetone, and alcohol. Not for use with hydrocarbons.	300°F (150°C)	BA, CA, DA
FKM: Fluorocarbon (Viton®)	Water, saltwater, wastewater, most chemicals and solvents, aromatic hydrocarbons and fuels, chlorinated hydrocarbons, and high temperatures.	390°F (200°C)	HK
IIR: Isobutene isoprene (Butyl)	Water, saltwater, wastewater, low permeability, and outdoor service. Not for use with hydrocarbons.	230°F (110°C)	AA, BA
NBR: Acrylonitrile butadiene (Nitrile, Buna-N)	Water, saltwater, wastewater, refined petroleum hydrocarbons, chlorinated solvents, fats, and greases.	250°F (120°C)	BF, BG, BK, CH
NR, IR: Polyisoprene (Natural rubber)	Water and saltwater. Not for use with residual disinfectants or hydrocarbons and chlorinated solvents.	212°F (100°C)	AA
SBR: Styrene butadiene	Water, saltwater, and wastewater. Not for use with hydrocarbons and chlorinated solvents.	212°F (100°C)	AA, BA
VMQ: Polysiloxane (Silicone)	Water, saltwater, wastewater, outdoor service, hot air, and electrical insulation. Not for use with hydrocarbons and chlorinated solvents.	420°F (215°C)	FC, FE, GE

* Water (i.e., source, potable, or reclaimed wastewater); wastewater (i.e., a combination of used liquid and liquid-carried waste from residences, commercial buildings, industrial plants, and institutions, which may include the inflow of groundwater, surface water, and stormwater).

Courtesy of Parker Hannifin Corporation and Pruet.

may be damaged if it is subject to significant swelling in service. A valve diaphragm must retain its strength and strain ability throughout its life.

The best indicator of performance is experience if in situ components can or have been monitored. If a particular water distribution system has operated for decades with particular elastomeric materials in certain applications, that knowledge should be documented and used to develop future requirements.

When selecting elastomers for use in the water industry, resistance to certain chemicals, especially chlorine and chloramines, must be considered. ASTM publishes standard D6284, which gives test protocols at various temperatures and concentrations of chlorine and chloramines but offers no acceptance criteria (also see Chapter 5, Elastomer Degradation Trends).

A purchaser or designer of valves, pipes, or fittings with knowledge of rubber compounds can better understand and discuss compounds with the rubber supplier, noting the particular needs of a specific application. Is the seal static or dynamic and subjected to what pressures and forces? Will the seal be subjected to compression, tension, or shear and at what levels? What percentage of the seal is exposed to the fluid? The answer to these questions will help in the evaluation and selection of elastomers or, if necessary, the development of an elastomer formulation.

A general listing of common rubber materials used for seals and gaskets in the waterworks industry is shown in Table 3-3. These are general guidelines for elastomers, which are greatly affected by other ingredients in the formulation. Moreover, temperature ratings vary by whether the application temperature is dry air, hot water, or steam. In general, elevated temperatures cause additional vulcanization, which can make the elastomer harder and less resilient. Hence, elastomers may be acceptable at the stated temperatures for intermittent exposure; however, long-term exposure may damage the seals or shorten their life.

These guidelines only serve as a general starting point. Chemical resistance guides (Chapter 5) and gasket design recommendations from suppliers should also be used for the final selection. Finally, there is no substitute for the operating experience of a specific elastomer in the intended or a similar service.

MANUFACTURE OF SEALS AND GASKETS

Once an elastomeric compound is selected, the seal or equipment manufacturer must use a documented process for mixing, molding, and finishing the product. The first step in the process is to mix the ingredients at a specified temperature in the exact quantities in a rubber mill or internal mixer. In a rubber mill, the base elastomer and the additives are loaded between two rollers and pressed into a homogeneous mixture in about 30 min (Figure 3-1). Alternatively, a closed internal mixing system can be employed where the ingredients are fed into a closed chamber and mixed by rotors in about 5 min. Currently, mixing of rubber compounds is predominantly done using a closed mixing process. A closed mixing process provides better compound homogeneity, a shorter mixing process, and improved occupational health and safety. The output of any mixing process is a long ribbon or thick sheet of uncured rubber with the consistency of bubblegum. The material can be stretched, but it has no resiliency until vulcanized at high temperature.

One manufacturing process used to make gaskets and diaphragms is the calendar process wherein rollers are used to precisely control the thickness of the finished rubber sheet. The flat material is then slowly pulled through a heated die where it is thermally cured for 20–30 min. The cured sheets can then be die-cut into various shapes for gaskets or valve diaphragms.

Courtesy of AVK Gummi A/S.

Figure 3-1 Rubber ingredients are loaded between two rollers for mixing

Figure 3-2 Rubber lining on the interior of a butterfly valve body

If uncured rubber is used as a lining material, it is rolled onto the metal surfaces precoated with a bonding agent or adhesive at room temperature, and then cured in a steam-heated autoclave until the rubber is fully bonded and cured to the pipe, valve, or fitting. Figure 3-2 illustrates the interior of an NPS 24 (DN 600) gray iron butterfly valve body with ⅛-in. (3.2-mm)–thick rubber lining for improved corrosion resistance after removal from the autoclave.

The most common rubber forming processes are compression, transfer, and injection molding. A hardened steel mold is typically used, which has a cavity or multiple cavities

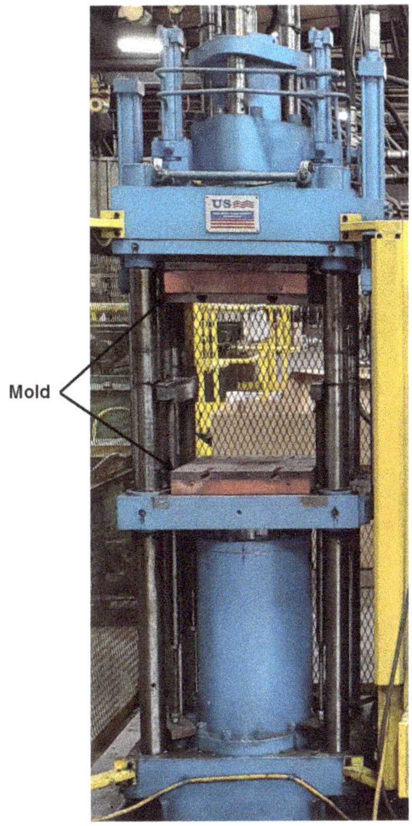

Courtesy of Specification Rubber Products, Inc.

Figure 3-3 Typical rubber molding press with electric heating

in the shape of the part to be molded. The mold is mounted into a rubber press and heated with electrical current or steam to about 149°C (300°F) (see Figure 3-3). The mold cavity is sprayed with a mold release agent to prevent the part from adhering to the mold during vulcanization and allow easy removal from the mold. Some molds have the capability of supporting production parts such that metal parts can be encapsulated by the rubber. Inserted metal parts are cleaned and coated with a bonding agent prior to assembly of the mold.

In compression molding, uncured rubber pieces in the form of premeasured slabs or ribbons are placed in the mold cavity and pressed at very high loads (i.e., tons) for about 20 min at a high temperature while the rubber vulcanizes. With the transfer mold process, a charge of rubber is injected into the heated mold cavity with a plunger (see Figure 3-4). Finally, with an injection mold process, the uncured rubber flows into the cavity under pressure while the mold is pressed (see Figure 3-5). Injection molding provides parts with better tolerances, higher curing degree, and hence, better part performance (due to higher degree of cross-linking).

After the molding time is complete, the mold is opened, and the fully cured part is removed. Certain elastomers may also go through a post-cure process in an autoclave. Any flashing is removed with an abrasive tool or sharp blade, and the part is ready for use (see Figure 3-6). The molding process is repeated until the desired number of parts is produced.

Some valves have a molded-in seat (see Figure 2-19). In these valves, the valve body becomes part of the mold, and a mold insert coated with mold release agent is used for the opposite side of the rubber seat. In the case where the rubber must adhere to the valve

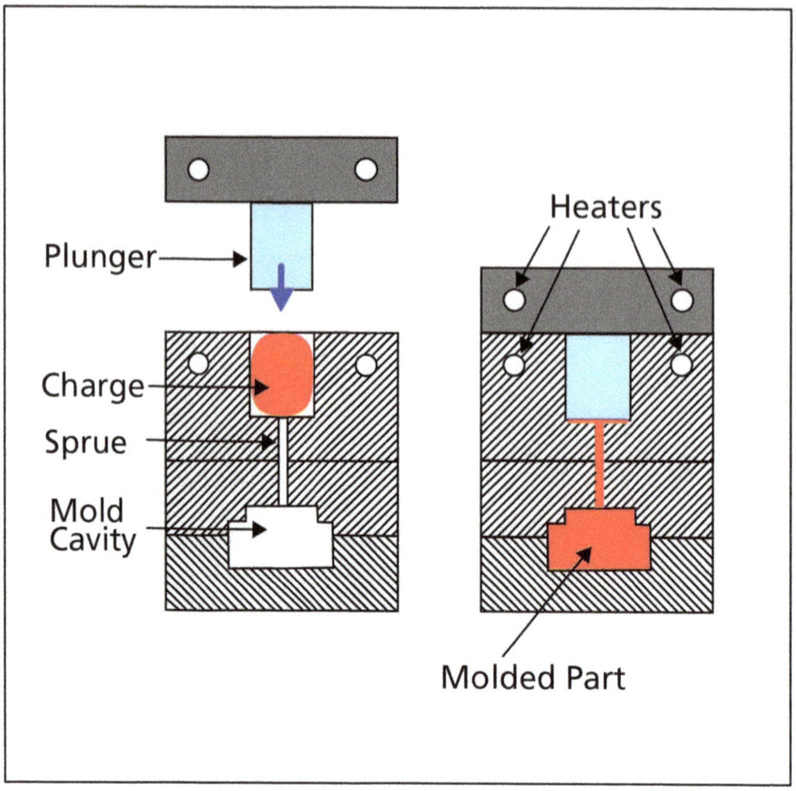

Figure 3-4 Typical transfer molding process

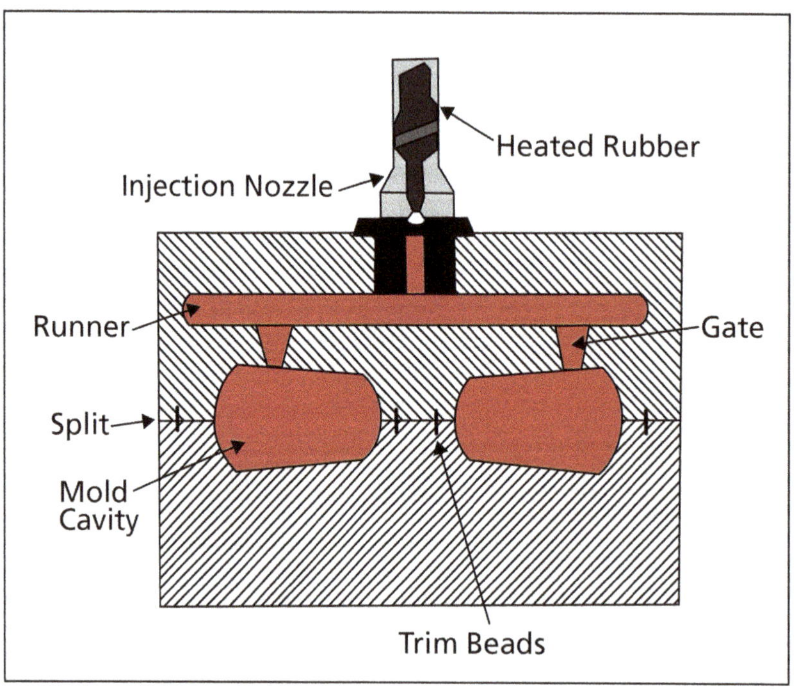

Figure 3-5 Typical injection molding process

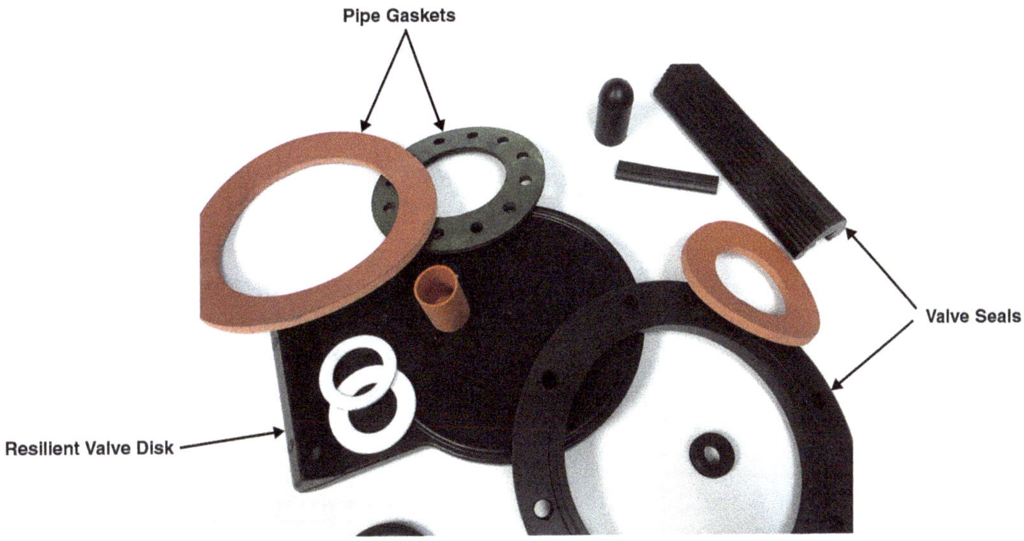

Figure 3-6 Samples of molded rubber parts

Courtesy of Specification Rubber Products.

Figure 3-7 Fabrication of seals using an extrusion die

body or gate, the metal surface of the valve must be sand blasted to provide an adequate rough surface profile to assist in mechanically anchoring the rubber to the valve body's surface. A primer and adhesive or bonding agent is then applied to the valve surfaces to chemically anchor the rubber to the valve's body but not to the mating mold. The raw elastomer compound is inserted or injected into the cavity and held at a high temperature for a specified time period to allow vulcanization to occur. Finally, the mold is opened, and the finished, fully cured molded part is removed.

Another variation of the injection molding process is to extrude the rubber through a die (Figure 3-7) and then post-cure the rubber in a continuously fed heated oven.

It is common for large-diameter O-rings and pipe gaskets to be fabricated in the field by splicing extruded round cord stock or molded lengths of a seal. The process requires great care and consists of the following steps (Rahman 2007):

- Cut the ends of the length perfectly square.
- Apply a splicing adhesive for the specific rubber compound.
- Place the end sections in a splice mold, and apply heat and pressure for several minutes.

The integrity of the splice can be verified by stretching the gasket to at least twice its original length while rotating a minimum of 180° in each direction to allow inspection of the splice. Any visible separation or cracks shall be cause for rejection (AWWA 2017).

MANUFACTURING CONSIDERATIONS

Regardless of the method of producing rubber parts, there are several important considerations when employing these processes. Rubber is not rigid and is subject to excessive hardening, shrinkage, and permanent deformation during the molding process. Therefore, controlling precise tolerances in its production is difficult. The surface finish of the parts is checked for blistering or cracking, which could be a result of a variation in the molding process, such as mold temperature. Dimensional variations may result from insufficient mold pressure or excessive injection pressure.

All molding processes should be accompanied with a robust quality control program. Each batch of parts should be checked for rubber hardness by hand or an automated machine, which will indicate if the curing process was sound. The rubber compound can also be checked with a rheometer test machine (see Figure 3-8), which records the viscosity during the vulcanization process of a sample test specimen. The process makes use of two opposing dies, which oscillate while the torque is measured. Rheometers are often used in R&D labs for developing new compounds.

Courtesy of Specification Rubber Products.

Figure 3-8 Rheometer fixture

Once parts are complete and approved, they should be stored in a cool, dry, and dark environment. Excessive heat will cause parts to overcure and harden. Moist or unclean locations can cause the parts to take on moisture or contamination. UV sunlight or exposure to ozone from electrical equipment can cause surface cracking and loss of strength. When stored properly, parts will remain like new for more than five years.

Because of its elasticity, a rubber part has the advantage of being forgiving in application. Rubber properties are verified using test slabs with a specific set of dimensions that can vary greatly from the finished part. As a result, the as-molded properties of the part in terms of strength or elasticity may vary slightly, which could affect performance. Finally, when rubber parts are used, they are often lubricated with special lubricants and greases that can also affect their performance.

Many AWWA valve and fitting standards contain proof-of-design and production test protocols to verify the performance of the rubber parts in their intended service. In addition, certain aging and environmental tests, such as hot air aging and ozone exposure, are also used in the standards to assure the performance of the material in service over its intended life.

REFERENCES

Allstate Gasket. 2007. *Gasket Specifications Guide.* www.allstategasket.com/info_specs_material-properties.asp (accessed January 2007).

American Society for Testing and Materials (ASTM) International. 2017. ASTM D2000, Standard Classification System for Rubber Products in Automotive Applications. Philadelphia, Pa.: ASTM International.

ASTM International. 2017. ASTM D6284, Standard Test Method for Rubber Property-Effect of Aqueous Solutions with Available Chlorine and Chloramine. Philadelphia, Pa.: ASTM International.

American Water Works Association (AWWA). 2017. AWWA C300, Reinforced Concrete Pressure Pipe, Steel Cylinder Type. Denver, Colo.: AWWA.

Blair Rubber Company. 2018. *Blair Engineering Manual.* blairrubber.com/engineering-manual/.

Harper, C.A. 1992 (2nd ed.). *Handbook of Plastics, Elastomers, and Composites.* New York: McGraw-Hill.

Parker Hannifin Corp. 2018. *Parker O-Ring Handbook ORD 5700.* www.parker.com/literature/ORD%205700%20Parker_O-Ring_Handbook.pdf.

Pruett, K.M. 1994 (2nd ed.). *Chemical Resistance Guide for Elastomers.* California: Compass Publications.

Rahman, S. 2007. Sealing Our Buried Lifelines. *AWWA OPFLOW*, April.

Reiber, S. 1993. Investigating the Effects of Chloramines on Elastomer Degradation. *Journal AWWA*, August.

Rockaway, T.D.; A Willing, G.A.; Schreck, R.M.; and Davis, K.R. 2007. *Performance of Elastomeric Components in Contact with Potable Water.* Denver, Colo.: AWWA Research Foundation.

This page intentionally blank.

Chapter 4

Mechanical Properties of Elastomers

The design of most valves, fittings, pipe joints, and other components of water systems include elastomeric components for resilient sealing. Users of these products should consider the given mechanical properties of the specified elastomer in the product's intended service. Elastomers are formulated from base polymers made of long chains of molecules, which are cross-linked to adjacent chains during the manufacturing process to form a compound. Typically, a compound is cured for 15–30 min within a temperature range of 300–400°F (150–200°C). The resultant cross-linking that occurs from the curing process provides the elastomers with their elasticity, and their resilience (ability to recover their original shape), once the load is removed. The strength and resilience of the chains give elastomers their longevity.

This chapter will present the composition and mechanical properties of elastomers typically used in the waterworks industry. The chapter that follows will address the degradation of these properties over time in various chemical and environmental exposures.

COMPOSITION

Elastomers are composed of atoms, typically carbon, hydrogen, and oxygen, that form long chain molecules. These long chain molecules are vulcanized with sulfur or peroxide to cross-link the chains randomly with each other to give the elastomers strength, resilience, and toughness.

Figure 4-1 illustrates the form of a single monomer unit for natural rubber as derived from natural latex, which is generally obtained from Pará rubber trees. The "n" suffix indicates that the molecule is repeated into a long chain of hundreds of thousands of units, which creates the natural rubber polymer and provides for its flexibility.

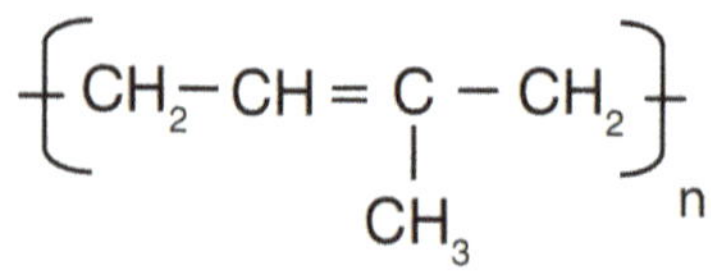

Figure 4-1 Single monomer of natural rubber

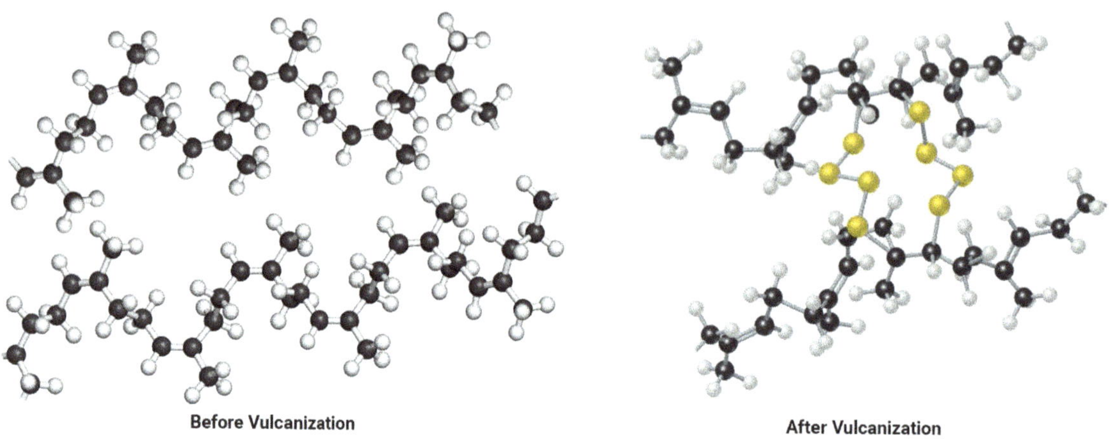

Source: StudioMolekuul/shutterstock.com. Reprinted by permission of Pearson Education, Inc.

Figure 4-2 Vulcanization of natural rubber with sulfur

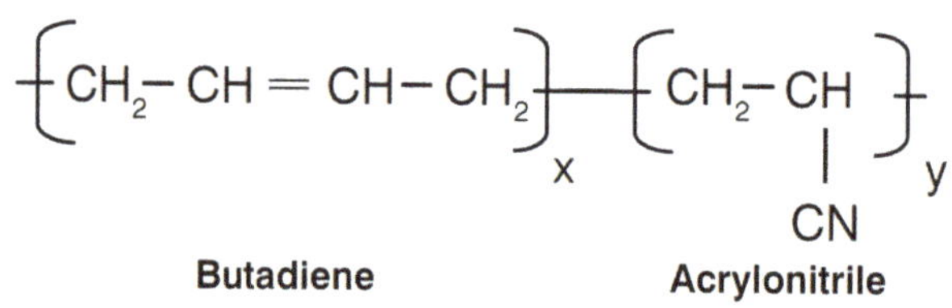

Figure 4-3 Composition of NBR (nitrile)

An illustration of an elastomer before and after cross-linking or polymerization with sulfur under heat and pressure is shown in Figure 4-2. The cross-linking anchors the chains of monomers at random points so that instead of sliding past each other, they provide elasticity and strength.

An example of a common synthetic polymer is nitrile (NBR) (Figure 4-3). Nitrile consists of a chain of two monomers, butadiene and acrylonitrile. Nitrile is a commonly used sealing material and displays a strong resistance to exposure from oil and fuel. This polymer has an "x" suffix and a "y" suffix, indicating that its properties can be varied by changing the ratio of the two monomers.

All base polymers are flexible and relatively impermeable to water, but their mechanical properties can vary widely. The selection of the base polymer is the most important decision in obtaining the desired elastomer's mechanical strength and chemical resistance.

The general properties of elastomers can be modified through the selection of various compounding ingredients. Some compounding ingredients are added to accelerate the vulcanization process, while others enhance the properties of the finished material or extend its usable life. Some of the common ingredients and their functions are presented in Table 4-1.

The majority of elastomers are vulcanized with the vulcanizing agent sulfur. Organic peroxides and metallic oxides can also be used to vulcanize certain polymers to provide greater chemical and oxidation resistance.

Fillers are used to promote the manufacturing and molding process and provide additional mass to the polymer that can be as much as 60 percent of its total weight. The filler selected and its percentage of polymer weight can affect the hardness and smoothness of the elastomer. The most common filler is carbon black, which is added as a fine powder in its elemental form and gives elastomers their black color.

Accelerants are chemicals added in small amounts to the polymer compound to speed up the curing process. Activators, such as stearic acid, are used in conjunction with the accelerant to increase its effectiveness.

Desiccants such as calcium oxide can be added to help remove moisture during the curing process.

Pigments are typically dry colored powders added to provide color to the elastomer.

Plasticizers are added to the polymer compound to increase its workability and flexibility.

Additives, such as antidegradants or antioxidants, are added to the polymer compound to improve its resistance to chemical attack or environmental attack from oxygen, ozone, and ultraviolet light. They diffuse into the elastomer and react with the chemical oxidant before it can damage the polymeric structure. These additives can greatly extend the useful life of an elastomer in a known environment.

SAFETY

Elastomeric components are widely used in drinking water systems and can contribute contaminants to the water. A complex set of tests and approvals have been put in place

Table 4-1 Common ingredients of elastomers and their functions

Ingredient	Name	Function
Vulcanizing agent	Sulfur, peroxide	Primary curing chemicals for cross-linking.
Fillers	Carbon black, clay, talc, silica	Provide reinforcement and accelerate cure. Super-fine clays can improve mechanical strength. Help lower cost.
Accelerant	Guanidine, thiazole, sulfenamide, thiuramdisulfide	Control the cure speed of the material.
Activators	Zinc oxide stearic acid	Initiate the curing process.
Desiccants	Calcium oxide	Remove moisture during the curing process.
Pigments	Disazopyrazolone	Give polymer a hue.
Plasticizers	Adipate	Reduce hardness and help incorporate the filler material.
Additives	Antioxidants, extending oil, antiozonants	Improve properties of the compound such as aging.

to protect consumers from possible contaminants based on the consensus standard from NSF International, NSF/ANSI/CAN 61, Drinking Water System Components—Health Effects. This standard was developed to prevent potentially harmful health effects from the leaching of contaminates or impurities from materials coming in contact with drinking water, including process media, elastomers, coatings, and mechanical devices. NSF/ANSI/CAN 61 does not cover product performance; it only seeks to provide criteria for evaluating contaminants and impurities imparted indirectly to drinking water.

Various certification organizations may be involved in certifying products in accordance with NSF/ANSI/CAN 61. Certification organizations are accredited by the American National Standards Institute (ANSI) and currently include NSF International, Truesdail Laboratories, the Canadian Standards Association (CSA) Group, Underwriters Laboratories, the Water Quality Association, and the International Association of Plumbing and Mechanical Officials (IAPMO) Research and Testing. Individual states or local agencies have authority to accept or accredit certification organizations within their jurisdictions. Currently, 49 states and Canada mandate its use. Accreditation of certification organizations may vary from jurisdiction to jurisdiction. Waterworks product standards do not typically mandate the certification of products, except for potable water use.

The NSF/ANSI/CAN 61 certification process begins with the manufacturer providing the certification organization with a wetted parts list identifying the materials of construction of the product coming in contact with drinking water. Coatings and elastomers are further defined with a list of its components or ingredients down to the specific chemicals and elements in the material. Products and coupons are then submitted for evaluation.

The certification lab conditions the samples in 73°F (23°C) chlorinated water for 14 days. Where hot water applications are required, the test water temperature is elevated accordingly. The samples are then exposed to test waters at various pH levels for 19 days, and water is extracted at prescribed intervals during the testing period for analysis. The test water is analyzed based on the expected contaminants derived from the wetted parts lists and the list of ingredients. The results of the sampling analyses display the known contaminants being quantified in mg/L. The results are multiplied by a normalization factor that accounts for differences in surface area–to–volume ratios between the sample and actual in-service conditions. For example, an elastomer in a valve may be tested in a liter of water, but in-service, it may be exposed to thousands of gallons of water under flowing conditions, thereby contributing far less contaminants to the drinking water system than those indicated by the test results. Finally, the results are compared to established criteria for maximum contaminant levels (MCL), and the sample receives a pass/fail result.

The NSF/ANSI/CAN 61 certification process is product-based and specific to each manufacturer, so if an ethylene propylene diene monomer (EPDM) gasket is certified for one manufacturer for a specific end product, a similar EPDM gasket from another gasket manufacturer would not be certified for the same end product. Product listings are presented on the certifier's website for public reference (see sample in Figure 4-4).

STRENGTH

Elastomeric components should be engineered to resist excessive distortion, tear, aging, and permanent set while subjected to various pressure, compression, and shear load requirements of the application. Once the seal designer quantifies the load and stress condition, the characteristics of various elastomers can be evaluated. Elastomers can be formulated from a long list of base polymers as listed in Table 3-3 and various additives and chemicals as listed in Table 4-1, which are cured to varying process parameters to achieve the desired strength requirements. The resultant mechanical strength, elasticity, and chemical stability can vary widely among elastomers. To assist the designer in selecting

Water Quality Association

10/9/2018

CERTIFIED DRINKING WATER SYSTEM COMPONENTS

NSF/ANSI 61 - 2016: Drinking Water System Components - Health Effects

Company: ABC Corp.

Product Type: Valve

Brand Name	Model	Water Contact Temperature	Water Contact Material	Size
ABC Check Valve	1234.XX	Domestic Hot (60C)	Various	Minimum 2"

Adapted from Water Quality Association.

Figure 4-4 Sample NSF/ANSI/CAN 61 online certification

Table 4-2 Required properties for a push-on pipe gasket

Mechanical Property	Requirement	Test Method
Tensile strength, min.	2,000 psi (13.8 MPa)	ASTM D412
Ultimate elongation, min	400%	ASTM D412
Compression set, max.	20%	ASTM D395, Method B
Hardness Type A durometer	40–60	ASTM D2240

Adapted from ASTM F477.

a compound, several standardized tests have been developed to specify and measure the various strength properties of elastomers. As an example, the mechanical properties of pipe gaskets are given in Table 4-2. The mechanical properties of elastomeric components vary widely depending on the selected compound and function.

The "hardness" of an elastomeric compound is related to its strength and can be measured by a durometer instrument, which uses a calibrated spring-loaded pointer to measure the depth of penetration when the device is momentarily pressed against a sample of the elastomer (ASTM D2240).

There are three durometer devices measuring hardness on three scales (00, A, and D). The Shore A durometer device, shown in Figure 4-5, is used for a range of soft to hard elastomers and is calibrated to indicate a durometer of 100 when pressed against a hard glass surface.

Shore 00 durometer devices and scales are available for extremely soft elastomers, and Shore D durometers are suited for very hard elastomers and plastics. For example, a household rubber band will display a hardness of 20 Shore A, while the rubber in an automobile tire tread may exhibit a 70 Shore A hardness (see Figure 4-6). When elastomer samples are measured with these devices, the results will vary by a few points based on the technique of the user. Also, because the compound properties vary by mix and batch,

Figure 4-5 Durometer device (A scale)

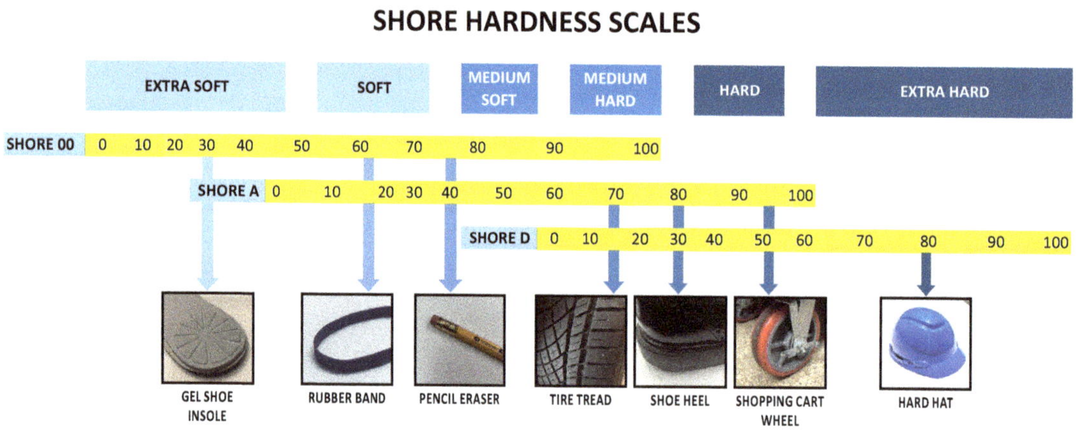

Figure 4-6 Various durometer scales

a durometer measurement is usually specified with a plus or minus tolerance of 5 points (e.g., Shore A 60 ±5).

A low durometer seal will be very flexible under load and more easily conform to the surfaces that it contacts. Hence, if a designer needs a component to seal well at low pressures such as 20 psig (138 kPa), an elastomer with a low durometer (i.e., Shore A 30) may be specified. Conversely, when a component must seal high pressures such as 500 psig (3,450 kPa), a high durometer (i.e., Shore A 80 or 90) may be specified. Through compounding of ingredients, all elastomers can be manufactured to a wide range of hardness.

A related measurement of elastomer strength is its "tensile strength" measured in pounds per square inch (psi) or megapascals (MPa) required to rupture a specimen in tension in accordance with an industry test standard, such as ASTM D412. To determine the tensile strength of an elastomer, a precise die is used to cut a specimen with an approximate length of 4.5 in. (114.3 mm) from a cured elastomer test slab with an approximate thickness of ⅛ in. (3.2 mm) as shown in Figure 4-7. The specimens are referred to as rubber dumbbells because of their appearance.

The specimen is placed in a tension testing machine as shown in Figure 4-8 and supported by two cylindrical knurled grips or clamps. A motor-driven screw rod raises the

Figure 4-7 Elastomer test specimen

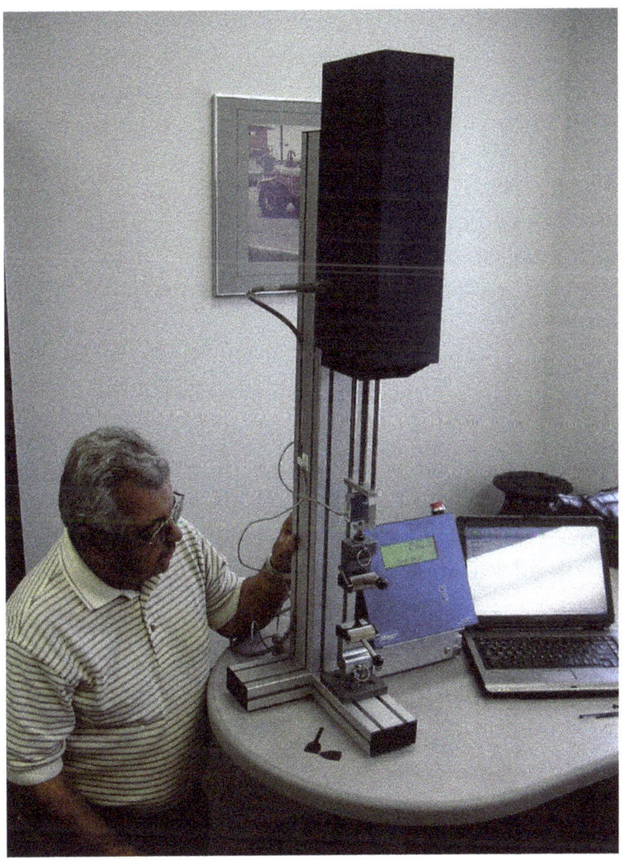

Figure 4-8 Tension testing machine

upper clamp and stretches the specimen at a uniform rate of 20 in. (508 mm) per minute until the specimen ruptures. The measured load at rupture divided by the cross-sectional area of the unstretched center portion of the specimen is reported as the tensile strength. The tensile strength of a typical elastomer used for a 70 Shore A durometer O-ring may typically range between 1,000 and 2,000 psi (6.9 and 13.8 MPa). The tensile strength is often used in seal design applications because the expected tensile stress can be calculated by design engineers for valve components such as pressure diaphragms.

During the tension test, the specimen will stretch to several times its original length before rupture, not unlike a rubber band. To measure the amount of stretch, two parallel benchmarks are made across the reduced section of the specimen at a known distance prior to the test. Then, during the test, the distance between the two marks is measured and the maximum distance is recorded. The "elongation" of the material is calculated by the following formula:

$$E = 100 \times (L - L_o) / L_o$$

where:

E = elongation, percentage
L = measured length between marks at rupture, in. (mm)
L_o = original length between marks, in. (mm)

Elongation is a measurement of the elasticity of the elastomer. A higher elongation represents a material with greater resilience and forgiveness. A rubber band will likely have an elongation of over 600 percent. A valve or fitting gasket may have a lower elongation, such as 300 percent. In general, the higher the tensile strength of the elastomer, the lower the elongation will be. A high-pressure O-ring needs to have a high tensile strength so that it does not distort and displace under pressure. Because it is usually contained in a sealing groove, it does not require a high degree of elasticity (i.e., elongation) to create a seal. On the other hand, a valve diaphragm may need high flexibility (i.e., elongation) to allow the movement of valve mechanisms during operation.

It is important to note that tensile strength, as well as other elastomer characteristics, generally change over time. Many kinds of exposure testing can be used to predict the reduction in tensile strength. Exposure to elevated temperatures, large loads, or deleterious concentrations of chemicals applied in a controlled method can provide reliable strength reduction results.

RESILIENCE AND TEAR RESISTANCE

Like elongation, resilience is a measure of the flexibility of an elastomer but, more specifically, the elastomer's ability to return to its original shape after a load is removed. When a rubber gasket is squeezed between two flanges, as shown in Figure 2-5, the reduction in thickness of the gasket results in changes to its inside and outside diameters so that the overall volume remains the same. Then, once the load is removed, the gasket recovers by returning back to its original thickness and diameter. Thus, elastomers are said to deform, not compress.

One exception to this behavior is an elastomer used as a cushioning material, which will have closed cells in its structure containing compressible air. There may be some measurable permanent deformation, but it may not be noticeable or affect the part's function.

One property that is used to measure the permanent deformation and the recovery of an elastomer is called compression set, which is typically conducted in accordance with ASTM D395, Method B. To conduct the test, a precise die is used to cut a round specimen

Courtesy of Specification Rubber Products, Inc.

Figure 4-9 Compression set apparatus

with a diameter of 0.5 in. (13.0 mm) from a cured elastomer test slab with a thickness of ¼ in. (6.0 mm). Several specimens are deflected between steel plates with fixed-height spacers as shown in Figure 4-9.

During the test, the samples are deflected approximately 25 percent for a fixed period, such as 22 h, at a constant temperature, and allowed to recover over a specified period (e.g., 30 min) at room temperature. Compression set is calculated as the percentage of the original deformation not recovered using the following formula:

$$C = 100 \times (t_o - t_i) / (t_o - t_n)$$

where:

C = compression set, percentage of original deflection
t_o = original thickness, in. (mm)
t_i = final thickness after recovery, in. (mm)
t_n = thickness under load, in. (mm)

A perfect elastomer demonstrating full recovery would have a compression set of 0 percent. Conversely, an elastomer that plastically deforms and exhibits no recovery would result in a compression set of 100 percent. Again, it should be noted that an elastomer is theoretically incompressible, so the sample permanently changed shape and not volume. In practice, the lower the long-term compression set (i.e., permanent deformation), the more resilient the elastomer. Product designers find the compression set an important parameter for seals because as valves open and close, the elastomeric seal often needs to deflect and recover to perform its intended function.

When seals are dynamic and subject to shear and tensile loads from moving parts, another parameter that can be used to evaluate the elastomer is the tear resistance measured in pounds per inch of thickness (lb-f/in.) or newtons per millimeter (N/mm). The measure of tear resistance is the force required to rupture a specimen in tension in accordance with an industry test standard, such as ASTM D624. This test requires a precise die to be used to cut a specimen similar to the tensile test dumbbell. Then, in the tear resistance test, the specimen will have a cut or have a 90-degree V-notch made prior to testing the sample as shown in Figure 4-10.

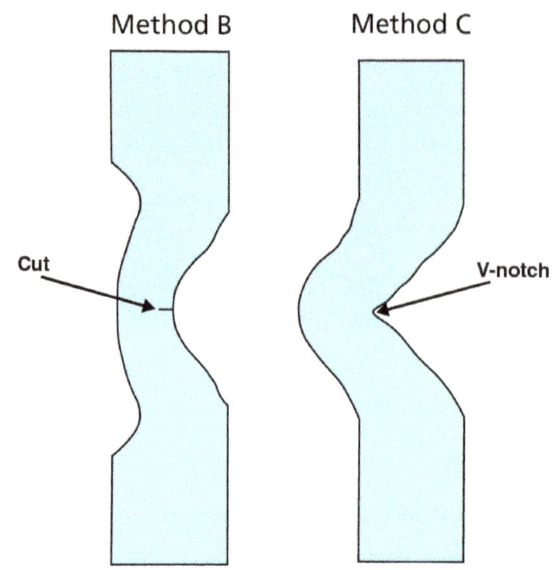

Figure 4-10 Tear resistance test specimens

TEMPERATURE RESISTANCE

The product design engineer must sometimes evaluate and select elastomers with respect to their hardness, tensile strength, elongation, and compression set at freezing or elevated temperatures. Valves in the water industry can be exposed to a wide range of operating environments and temperatures. For example, the AWWA C504 standard on butterfly valves lists a standard operating temperature range of 33–125°F (0–52°C). Hence, the designer must have an understanding of how temperature affects the properties of various elastomers across a range of temperatures. Conducting strength tests at elevated temperatures will reveal the effect of temperature on the selected elastomer.

In general, manufacturers of elastomers will publish a recommended operating temperature range as shown in Figure 4-11. While operating in the recommended temperature range, the properties of elastomers will vary. In cases where the elastomer is to be bonded to a metal surface, consideration to the properties of the bonding agents should also be evaluated.

Furthermore, as temperatures are elevated, the life of the elastomer may be reduced due to degradation. The change in strength over time at elevated temperatures is dependent on the material's chemical stability. In some cases, elevated temperatures can break the bonds or cross-links, which can result in softening of the material and a loss in strength. In other cases, an elevated temperature can produce additional cross-linking, which hardens the rubber and causes a loss of resilience. The rubber formulation and method of cross-linking determine which effect will apply under the given temperature environment.

On the other hand, testing at higher temperatures to accelerate the rate of elastomer degradation is a common procedure that provides useful information. Tests such as ASTM D573 take into account and adjust for the effect of temperature change on cross-link bonds.

Irrespective of the long-term temperature effects, the hardness of most elastomers will remain somewhat steady during the standard operating temperature range of 33–125°F (0–52°C). For nitrile and EPDM, the change in hardness between ambient and freezing temperatures is typically an increase of only 10–15 Shore A durometer (see Figure 4-12). For

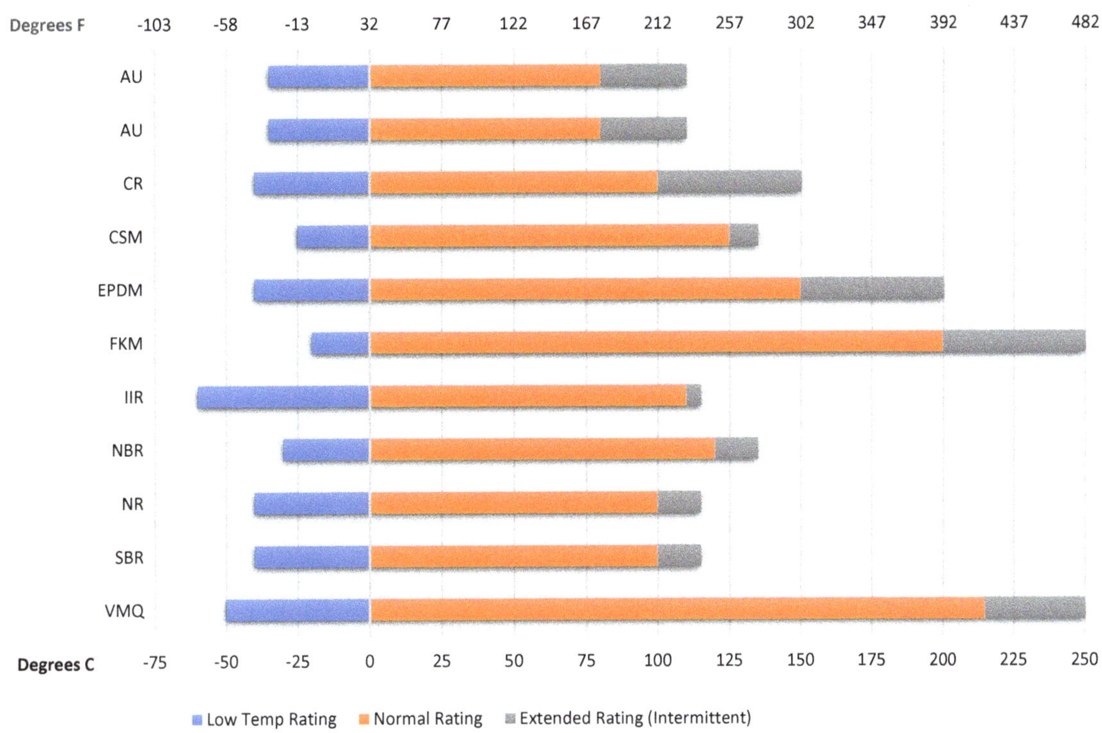

AU—polyester urethane, CR—chloroprene rubber, CSM—chlorosulfonated polyethylene, EPDM—ethylene propylene diene monomer, FKM—fluorocarbon, IIR—isobutene isoprene, NBR—acrylonitrile butadiene, NR—natural rubber, SBR—styrene butadiene, VMQ—polysiloxane

Figure 4-11 Typical operating temperature range of various elastomers

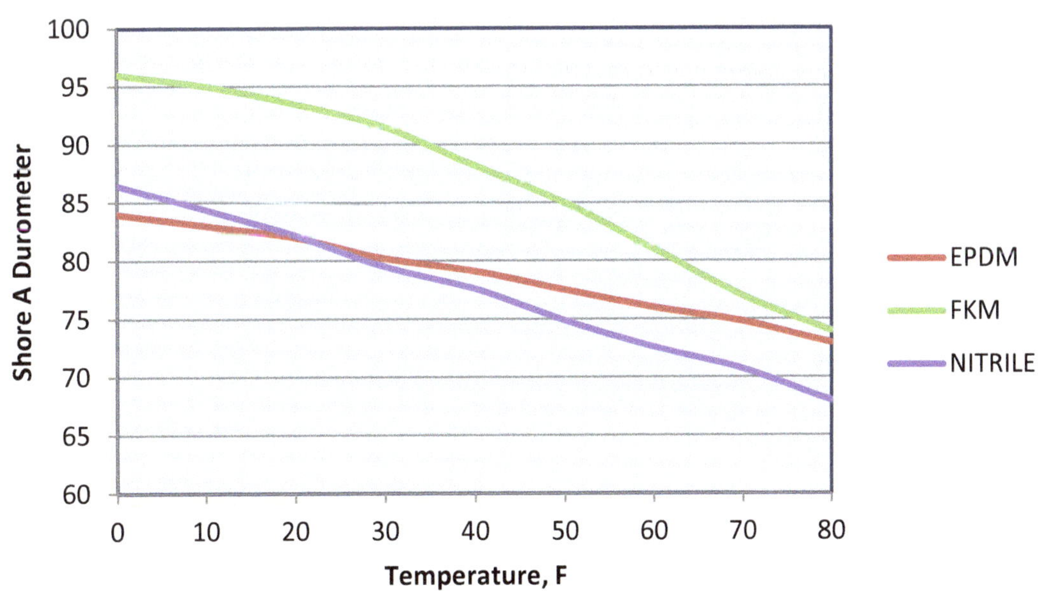

EPDM—ethylene propylene diene monomer, FKM—fluorocarbon

Figure 4-12 Durometer of elastomers at reduced temperatures

Table 4-3 Typical heat aging test variation for NBR (ASTM D2000)

Nitrile (BK) Heat Age, 70 Hours at 100°C	
Shore A durometer change, points	±10
Tensile strength change, % maximum	–20
Elongation change, % maximum	–30
EPDM (CA) Heat Age, 70 Hours at 125°C	
Shore A durometer change, points	+10
Tensile strength change, % maximum	–20
Elongation change, % maximum	–40

Adapted with permission from ASTM International.

FKM, the difference is more pronounced, and typically, a 20 Shore A durometer increase can be anticipated. This hardening may cause a seal to lose its flexibility and should be considered in the design of the seal. Conversely, at higher temperatures, a decrease of 10–20 Shore A durometer may occur, making seals more flexible and perhaps enhance their ability to seal.

One caution regarding elastomers at elevated temperatures, outside of ranges applicable to water treatment and distribution, is that a seal will grow significantly in size due to thermal expansion. Because the thermal expansion of rubber can be as high as 10 times that of steel, the space within which a seal resides should be designed to accommodate the increased volume of the seal at the expected maximum temperature.

The compression set is adversely affected by increased temperature as most elastomers tend to become more plastic at elevated temperatures. One way to predict the resistance of elastomers to temperature exposure is to perform an accelerated hot air aging test. As an example, when NBR and EPDM are subjected to the ASTM D573 hot air oven test at 212°F (100°C) for 70 hours, the following change in properties can be expected as indicated in Table 4-3.

TEST PROTOCOLS

The designer should determine the required mechanical properties of the elastomeric component and review the reported properties of alternate elastomers that also provide the required chemical and environmental resistance. To assist in the evaluation, many standard test protocols are used to predict the performance of elastomers given in Appendix B and summarized in Table 4-4.

Table 4-4 Summary of standard elastomer tests

ASTM Elastomer Test Standards	
Percent ash, percent carbon black	ASTM D 297
Compression set method	ASTM D 395
Tensile, elongation and modulus properties	ASTM D 412
Adhesion to a flexible substrate	ASTM D 413
Adhesion to a rigid substrate	ASTM D 429
Chemical resistance	ASTM D 471
Deterioration in an air oven	ASTM D 573
Compression deflection	ASTM D 575
Tear resistance	ASTM D 624
Change in hardness by heating in air	ASTM D 865
Hardness (Durometer)	ASTM D 2240
Identification by infrared spectrophotometry	ASTM D 3677
Effect of chlorine and chloramine solutions	ASTM D 6284

REFERENCES

American Society for Testing and Materials (ASTM) International. 2016. ASTM D395, Standard Test Methods for Rubber Property–Compression Set. Philadelphia, Pa.: ASTM International.

ASTM D2240. 2015. Standard Test Method for Rubber Property—Durometer Hardness, ASTM International, West Conshohocken, PA.

ASTM International. 2016. ASTM D412, Standard Test Methods for Vulcanized Rubber and Thermoplastic Elastomers–Tension. Philadelphia, Pa.: ASTM International.

ASTM International. 2015. ASTM D573, Standard Test Methods for Rubber–Deterioration in an Air Oven. Philadelphia, Pa.: ASTM International.

ASTM International. 2012. ASTM D624, Standard Test Methods for Tear Strength of Conventional Vulcanized Rubber and Thermoplastic Elastomers. Philadelphia, Pa.: ASTM International.

ASTM International. 2017. ASTM D2000, Standard Classification System for Rubber Products in Automotive Applications. Philadelphia, Pa.: ASTM International.

ASTM F477. 2014. Standard Specification for Elastomeric Seals (Gaskets) for Joining Plastic Pipe, ASTM International, West Conshohocken, PA.

American Water Works Association (AWWA). 2015. ANSI/AWWA C504, Rubber-Seated Butterfly Valves. Denver, Colo.: AWWA.

National Sanitation Foundation (NSF) International. 2018. NSF/ANSI/CAN 61, Drinking Water System Components–Health Effects. Ann Arbor, Mich.: NSF International.

Walker, J. 2012. *Elastomer Engineering Guide.* Surrey, United Kingdom.

This page intentionally blank.

Chapter 5

Chemical Resistance of Elastomers

ELASTOMER CHEMISTRY

Elastomeric materials are unique from many of the products common to the water industry. Rather than being formed of a crystalline molecular structure like metals, both synthetic and natural rubber elastomers consist of long, intertwined polymer chains that are intermittently bonded or cross-linked. The cross-links are small polymer chains that occur at regular intervals on the longer, backbone polymer chains that bridge or link the long chains to another adjacent long chain. These cross-links make the long chains of molecules act more as a group than as individuals, thus creating a flexible entanglement. Because of the lack of crystallinity, the macromolecular structure is stretchable under low stress and has a high degree of shape retention memory.

The type of cross-link bonds that are formed depends on the base polymer and on the curing agent. For natural rubber (NR), styrene butadiene (SBR), nitrile (NBR), and chloroprene (CR), the backbone polymer chains all contain double-bonded carbon atoms that are available to form cross-links during curing. However, not all of the available bonding sites form cross-linkages, and ultimately, only a small number of cross-linkages are completed. Thus, elastomer degradation occurs when chemicals in the aqueous environment attack both the weak cross-link bonds and the available carbon sites (i.e., open sulfur bond locations) on the backbone, or the long polymer chain. In addition to removing the cross-linkages, the long, backbone chain length can become severed, which further reduces polymer entanglement to the point that parts of the network break away (see Figure 5-1).

The backbone molecular structure of ethylene propylene diene monomer (EPDM) is very different from that of other elastomeric materials. For EPDM, the double-bonded

Courtesy of J.D. Larson.

Figure 5-1 Elastomer degradation of a check valve disc in potable water service

carbon atoms available to form the cross-links are not contained within the polymer backbone chain. Rather, the double-bonded carbon atoms hang off the polymer chain as a pendant. Thus, the bonding sites available for cross-linkage are offset from the backbone (principal molecular structure of the long chain), and chemical or physical attack of the cross-linkage chain does not affect the base polymer backbone or the molecular long chain length. Therefore, the material properties derived from long chain entanglement will not be altered by chemical or physical attack.

As indicated, the curing agent used during the elastomer manufacturing process also has a significant effect on the formation of cross-link bonds and, thus, the elastomer's overall performance. Many natural rubber and synthetic elastomer materials are sulfur-cured. With sulfur curing, cross-links are formed when sulfur attaches to the double-bonded carbon in adjacent polymer chains. The cross-link can be short and of the form C–S–C, or relatively long and of the form $C–S_X–C$, where "x" is the number of sulfur nodes present in the cross-link. The longer cross-link chains have lower bond disassociation energy and, thus, are more susceptible to degradation (see Table 5-1).

Another commonly used elastomer curing agent is peroxide. While significantly more expensive than sulfur, the peroxide curing agent forces a C–C cross-link bridge between adjacent polymer chains. As the C–C bond has a much higher bond disassociation energy, the peroxide-cured elastomers are less susceptible to degradation.

The pendent double-bond carbon sites within EPDM that are available for cross-linkage can be formed using either sulfur or peroxide, which produce varying temperature and chemical resistance. Because EPDM contains relatively few cross-link sites compared to other elastomers, the material properties are primarily derived from the few cross-linkages and molecular entanglement. If cross-links are broken, the material properties

Table 5-1 Bond disassociation energy of common bonds

Bond Type	Bond Disassociation Energy (kJ $mole^{-1}$)
–C–C–	351
–C–S–C–	285
–C–S–S–C–	267
–C–S_X–C–	<267

Adapted with permission from Water Research Foundation.

are sustained by molecular entanglement. However, long-term material performance may suffer as entanglement is subject to thermodynamic creep and relaxation.

During peroxide curing of EPDM, peroxide is lost, depending on the amount of fillers and oils present, because of peroxide's immediate interactions with the acid surface of some fillers and with the small-molecular double bonds in some oils. To overcome this reaction, additional peroxide or cure time is used to have sufficient amounts available to form the cross-link chains. Therefore, there is typically more rubber included in EPDM-peroxide-cured compounds to increase the available cross-linkage sites and to form more carbon–carbon bonds (high disassociation energy). However, once the C–C site breaks, as noted previously, material properties are left to be dependent on polymer entanglement and, thus, subject to thermodynamic creep and relaxation.

In addition to selecting the base polymer material and curing agents, elastomer compounders tailor products to meet specific needs by adding vulcanizing agents, antidegradants, accelerators, retarders, pigments, plasticizers, processing aids, and fillers—all of which affect the cross-links. However, the exact formulations are proprietary and closely guarded.

DRINKING WATER COMPLIANCE

To promote public safety, the drinking water industry is heavily regulated. Many rules have been promulgated by the US Environmental Protection Agency (USEPA) to ensure the delivery of safe drinking water and prevent contamination. To help control biofilms and reduce coliform bacteria concentrations from forming, the USEPA mandates that a chlorine or chloramine residual be maintained within all points of the water supply's distribution system. Based on current regulations, the maximum residual disinfectant level (MRDL) for chlorine and chloramine is set at 4.0 mg/L by the USEPA. Compliance with the MRDL is based on a running annual arithmetic average, and water system may increase above 4.0 mg/L for emergency/short-term disinfections (https://www.epa.gov/dwreginfo/chloramines-drinking-water).

Over the past 100 years, free chlorine has been commonly used as a disinfectant in water distribution systems. It has proven effective against bacteria and waterborne microbes, with only a minor impact on water taste and odor. It is relatively inexpensive, readily available, and easily applied to water. However, recent work by the USEPA has indicated that prolonged exposure to trihalomethanes (THMs) from the chlorination process may lead to adverse health effects. To reduce the risk associated with THMs, many water distribution systems have switched from free chlorine to chloramines.

An alternative to chlorine disinfectant is the use of chloramines. Through the years, the prevalence of chloramine use in water treatment has varied. In 1936, 16 percent of US water treatment facilities used chloramines. The number declined in World War II because of a shortage of ammonia, which was needed for fertilizers, explosives, and other uses. The decline continued through the 1960s to a low of 2.6 percent (San Francisco Public Utilities Commission 2006). In response to the US Congress passing the Safe Drinking Water Act (SDWA) in 1974, many water distribution systems began switching their disinfectant from free chlorine to chloramines. By 1990, approximately 20 percent of treatment facilities were using chloramines. In 2002, USEPA data from 353 treatment plants showed that more than 46 percent were using chloramines. The USEPA estimates that the percentage of water treatment facilities using chloramines will likely increase to 60 percent as more stringent regulations gradually go into effect (Kirmeyer et al. 2004).

CHLORINE AND CHLORAMINE DISINFECTANTS

Drinking water disinfection commonly occurs through the addition of either chlorine or chloramine. Other methods of disinfection include ozone and ultraviolet (UV) light contained within special disinfection process systems. Further more, sections of the distribution system may be disinfected with a high dose (e.g., 200 mg/L) of bleach for four hours. Regardless of the disinfection process employed, the elastomers involved must be compatible with the process, oxidant, or disinfectant.

The two most common disinfection treatment processes are chlorine and chloramines. Chlorine has been used as a water disinfectant since the early 1900s. Chlorine and chloramines have adverse effects on some elastomer compounds in drinking water distribution systems.

Generally, when chlorine is added to water, the following reactions take place:

$$H_2O + Cl_2 \rightarrow HCl + HOCl$$
$$HOCl + H_2O \rightarrow H_3O^+ + OCl^-$$

The chlorine can be added to the water system as either gas or liquid in a pure form, or as a solid as either sodium hypochlorite, $NaClO$, or calcium hypochlorite, $Ca(ClO)_2$.

Chloramines, however, are produced when chlorine reacts with ammonia to produce a series of chlorinated ammonia compounds identified as either mono- (NH_2Cl), di- ($NHCl_2$), or tri-chloramine ($NHCl_3$). The type of chloramine formed depends on the pH and chlorine-to-ammonia ratio. At a chlorine-to-ammonia ratio of between 4 and 5 and at a pH of 8, monochloramine is the predominant chloramine. At a pH of between 5 and 8, both mono- and dichloramine form; at a pH of between 4.4 and 5, dichloramine is predominate; below a pH of 4.4, trichloramine is the predominate compound. Monochloramine (NH_2Cl) changes to dichloramine ($NHCl_2$) as the chlorine-to-ammonia ratio increases, then to trichloramine ($NHCl_3$), and eventually, to nitrogen gas or a variety of nitrogen-containing chlorine-free products. A simplified formation of the chloramines can be depicted as a step-by-step process (Sarai 2005):

$$NH_3(aq) + HOCl \rightarrow NH_2Cl + H_2O \text{ (Monochloramine)}$$

$$NH_2Cl + HOCl \rightarrow NHCl_2 + H_2O \text{ (Dichloramine)}$$

$$NHCl_2 + HOCl \rightarrow NCl_3 + H_2O \text{ (Trichloramine)}$$

In general, the water industry prefers monochloramine for disinfectant purposes because it has a longer residual life and has less impact on taste and odor than dichloramine or trichloramine.

ELASTOMER DEGRADATION TRENDS

It has been well documented that elastomeric components degrade when exposed to either chlorine or chloramine disinfectants. In the late 1980s and 1990s, there was increasing anecdotal information provided by water utilities that elastomeric materials were not performing well in chloraminated water distribution systems. However, the information assembled was unclear as some water systems that had been operating chloraminated systems for years reported no problems, while other systems that recently switched to a chloramine disinfectant were seeing degrading performance in some household and backflow prevention devices. Further investigation found damage to other water distribution appurtenances. Figure 5-2 shows damage to an encapsulated gate valve wedge.

To better understand the effects of chloramination on elastomer materials, the Water Research Foundation (formerly AwwaRF) initiated several research efforts. In 1993, Water Research Foundation commissioned a study on the *Chloramine Effects on Distribution System Materials* (Reiber 1993), in which 0.080-in. (4.4-mm)–thick elastomer coupons were exposed to 300 mg/L chlorine and chloramine at 100°F (38°C) for 30 days. Distribution system conditions are typically 3–5 mg/L and 73°F (23°C). The elevated concentration and temperature were needed to advance the aging and oxidation process to allow for reasonable observations in the laboratory. Various elastomer properties including maximum stress, swelling, and surface cracking were observed and documented.

The study by Reiber (1993) concluded that chloraminated compounds are significantly more aggressive to elastomers than chlorine, which would increase the potential for swelling, surface cracking, and loss of strength and elasticity. Testing of specific elastomeric compounds without antidegradants provided some general ranking of base elastomers for susceptibility to chloramine attack as shown in Table 5-2.

It is relevant to note that the elastomers tested were intentionally free of additives designed to prevent degradation in hopes of understanding the fundamental resistance of

Courtesy of J.D. Larson.

Figure 5-2 Degradation of a resilient gate valve wedge in service

Table 5-2 Susceptibility of elastomers without antidegradants to chloramine attack

Acutely Sensitive	Sensitive	Least Sensitive
Nitrile (NBR) Styrene butadiene (SBR) Natural rubber Neoprene	Ethylene propylene diene monomer (EPDM) Isobutyl isoprene	Fluorocarbon (FKM) Silicone

Source: Journal AWWA 1993.

the base polymer. But when commercially available sheet stocks were tested, they appeared to be more resistant to attack. Similarly, manufacturers now generally offer chloramine-resistant elastomers containing special antidegradants.

The study by Reiber (1993) caused a stir in the waterworks industry. Because nearly all US utilities possess thousands of buried elastomeric joint seals in their distribution systems made of acutely sensitive elastomers, a study was undertaken by the Ductile Iron Pipe Research Association (DIPRA) to investigate elastomer degradation in actual pipe gaskets versus thin elastomer test coupons because the use of test coupons does not represent the application-specific surface-to-volume ratio and product thickness of a pipe gasket. The study group tested 6-in. (150-mm) assembled push-on and mechanical-joint pipe assemblies and elastomer sheet specimens of sulfur-cured SBR, NBR, CR, EPDM, and FKM under the same conditions of 110 mg/L chloramine solution at 73°F (23°C) for 364 days. After the exposure, the FKM sheets showed little degradation while all of the other elastomeric compound sheets showed significant degradation. Conversely, none of the pipe gaskets showed any degradation after disassembly. Furthermore, the penetration of Cl_2 after 364 days was measured by X-ray dot mapping to be less than 1.3 mils (32.7 μm). The study concluded that testing of vulcanized sheet coupons is not relevant to predicting the performance of pipe gaskets (Bonds 2004).

The AWWA Committee on Gasket Materials has continued to study the use of elastomers in chloraminated water distribution systems. And, more recently, the AwwaRF report "Performance of Elastomeric Components in Contact with Potable Water" was completed in 2007, further expanding the understanding of potential elastomer degradation.

As part of these studies, an increase was observed with elastomer failures in water distributions systems that led to increased leaks, water main losses, unscheduled maintenance, and repair of components. The increased degradation appeared to coincide with the industry's disinfectant change from chlorine to chloramines. But other factors, such as temperature in warmer climates and pH, may also be contributing to increased elastomer degradation. The objective of all the noted research was to clarify the factors that contribute to elastomer degradation and develop a method for predicting the performance of elastomers in water distribution systems.

Accelerated-aging test protocols were evaluated with the goal of predicting long-term performance from the study's condensed period. Various elastomeric compounds were exposed to various ranges of temperature and concentrations of chlorine and chloramines for 30 days. As a result, life-cycle degradation curves of hardness, elongation, and volume swell were developed for generic formulations of eight commonly used elastomers.

The degradation results are valuable in illustrating the relative resistance to degradation of the generic compounds tested. But it is important to preface this information with the fact that the properties of a given elastomeric compound will vary widely with the addition of various additives and antidegradants.

As an example of the results of the 2007 AwwaRF study, Table 5-3 indicates how many years the particular laboratory formulation of the various generic compounds would resist a 50 percent swell in volume when exposed to water treated with 1 ppm chloramines.

Table 5-3 Degradation of various generic elastomers without antidegradants in chloramine service

Generic Elastomer	Years to Reach 50% Weight Change	Years to Reach 50% Strain Change
EPDM-peroxide	>30	>30
EPDM-sulfur	21.9	1.1
Nitrile	3.9	13.3
SBR	2.3	3.4
Natural rubber	1.9	17.4
Neoprene	0.4	1.8

Adapted with permission from AWWA Research Foundation 2007.
EPDM—ethylene propylene diene monomer, SBR—styrene butadiene

While the exact life span for a given compound is uncertain, product designers can see that, for this study, a generic peroxide-cured EPDM material is likely to outperform the other materials in water service with 1 ppm chloramine disinfection with consideration given to failure by swell and strain. However, if the particular part is not affected by these parameters, such as a pipe gasket, or various protective antidegradants are added, various other materials may be suitable as well (adapted from Water Research Foundation).

In addition, the study presented master degradation curves for various base elastomers with consideration given to maximum stress, maximum strain, and hardness for all three test concentrations. The degradation curves were based on the combination of tests at three distinct temperatures of 73°F, 113°F, and 158°F (23°C, 45°C, and 70°C, respectively) using the Arrhenius rate equation. The Arrhenius equation is a formula for the temperature dependence of reaction rates. A historically useful generalization supported by the equation is that, for many common chemical reactions at room temperature, the reaction rate doubles for every 10°C increase in temperature. The equation is widely used to develop accelerated-aging test protocols.

The application of degradation curves is dependent on the applicable failure mode of the elastomeric component, which may include loss of hardness, increase in volume or swell, and loss of tensile strength. The percentage of surface exposure and the elastomer's function must be considered when applying a prediction to an elastomer's degradation. For example, volume swell in a pipe gasket may not affect its intended function because it seals by compression. Conversely, volume swell of a butterfly valve seal may cause undesired additional interference and increased shear forces during closure of the valve. Also, some parts, such as push-on or mechanical joint pipe seals, have a very small area-to-volume exposure ratio. Hence, the degradation information is based on basic elastomers without antidegradants, and its use is dependent on the function of the component. In general, "the selection of the appropriate elastomeric material for a specific application must take into consideration the critical failure mode of the component" (Rockaway et al. 2007).

Even with the benefit of the chloramine studies of Reiber in 1993, Bonds in 2004, and Rockaway in 2007 as previously discussed, it is difficult to specify or qualify the correct elastomer for drinking water applications. To date, no standard chloramine test protocol in the waterworks industry exists nor are the critical failure modes of the various seals in the AWWA products identified. Work continues in several AWWA Standard Committees, and until such time as standard chloramine requirements are completed, the following informational note was prepared for waterworks products incorporating the use of elastomers in the foreword of the applicable AWWA Standards:

Chlorine and chloramine degradation of elastomers: The selection of materials is critical for water service and distribution piping in locations where there is a possibility that elastomers will be in contact with chlorine or chloramines. Documented research has shown that elastomers such as gaskets, seals, valve seats, and encapsulations may be degraded when exposed to chlorine or chloramines. The impact of degradation is a function of the type of elastomeric material, chemical concentration, contact surface area, elastomer cross section, and environmental conditions as well as temperature. Careful selection of and specifications for elastomeric materials and the specifics of their application for each water system component should be considered to provide long-term usefulness and minimum degradation (swelling, loss of elasticity, or softening) of the elastomer specified.

PERMEATION OF ELASTOMERS

Most drinking water distribution systems are buried and can be exposed to contaminated soil and groundwater containing hazardous chemicals, including petroleum products and low–molecular weight organic solvents. There have been more than 100 incidents of drinking water contamination in the United States, with the vast majority of them a result of leaking gasoline and heating fuel oil from underground storage tanks. Laboratory tests have shown that polyethylene (PE) pipe allows rapid permeation of gasoline and solvents and is not recommended in areas where contamination may be present. Polyvinyl chloride (PVC) pipe is impervious to gasoline exposure and other hazardous hydrocarbons at concentrations in water that are commonly encountered in the field (Ong et al. 2008). While PVC, iron, steel, and steel-lined concrete pipes can be impervious to permeation, their gasketed joints have been extensively studied (Glaza 1992) to determine their susceptibility to permeation.

Permeation is the process wherein a liquid or gas can pass through a substance on a molecular level as opposed to leakage, which involves the flow of fluids through an opening or gap. Glaza (1992) notes that drinking water contamination from gasket permeation may go unreported because the taste and odor threshold of users is not exceeded, yet a threat to health still exists.

Iron pipes can be joined by flanges (Figure 2-9), mechanical joints (Figure 2-13), or push-on joints (Figure 2-8). Plastic pipes can be solvent welded or joined by steel-reinforced push-on joints (Figure 2-7). Push-on joints have become the predominant gasket for water pipelines because of the ease in which the pipe can be assembled and the allowance for axial and angular deflection provided, which is needed for temperature changes and ground movements caused by seasonal changes and seismic events. These joints are also designed to prevent leakage for both positive and negative pressure events, which sometimes occur in pipelines and can cause infiltration of contaminated groundwater. SBR is the most common pipe gasket material, and it is estimated that it represents more than 90 percent of all installations in North America. NBR is sometimes specified when hydrocarbon resistance is important (Rahman 2007).

In permeation experiments (Cheng et al. 2012), ductile iron pipe sections with EPDM, CR, SBR, and NBR push-on gasketed joints were filled with water and immersed in various concentrations of premium gasoline. The interior water was analyzed for the aromatic contaminants from gasoline, including benzene, toluene, ethylbenzene, and xylene (BTEX). BTEX breakthrough times were found to be 35 days for EPDM, 50 days for CR and SBR, and 124 days for NBR. When a 100 percent aqueous solution of gasoline was tested, the breakthrough time was 210 days for SBR and 240 days for NBR and FKM.

Given that permeation will eventually occur through all pipe gaskets, the risk to drinking water quality must be assessed. Based on experimental BTEX permeation rates (Cheng et al. 2012), the permeation rate for various sizes of ductile iron pipe gaskets were extrapolated. It was estimated that the permeation rate of benzene from free gasoline exposure through a 4-in. (100-mm) gasketed joint was 6.56 mg/joint/day for SBR and 0.073 mg/joint/day for NBR. Given these rates, benzene contamination can be calculated for 8 h of stagnation (simulating an overnight scenario) in a length of pipe and for various water flows. The results show that the maximum contaminant level (MCL) for benzene, 5 μg/L, would be exceeded by a factor of 6 for SBR but acceptable for NBR. However, with typical water velocities in ductile iron water pipes at 2–10 fps (0.6–3 m/s), the benzene would be diluted to acceptable levels for both gasket materials and render the benzene undetectable.

In similar permeation experiments (Mao et al. 2011), 2-in. (50-mm) SBR and NBR Rieber (i.e., push-on) PVC pipe and gaskets were exposed to premium gasoline and saturated aqueous solutions of premium gasoline. When exposed to premium gasoline, BTEX breakthrough times were found to be as low as 21 days for the joints. Parallel tests confirmed no permeation through the PVC pipe material. Benzene permeation rates through the gasketed joints were found to be 0.73 mg/joint/day for SBR and 0.38 mg/joint/day for NBR. When exposed to saturated aqueous solutions of premium gasoline, no permeation through the pipe gaskets could be detected after 9 months of exposure. As with the ductile iron gasket study, for gasoline exposure, the benzene MCL will be exceeded for an 8-h stagnation period for 12 in. (300 mm) or smaller PVC pipes with SBR gaskets and 5 in. (125 mm) or smaller pipes with NBR gaskets. The MCL will not be exceeded with either gasket material when flowing water is considered. Finally, for gasoline-saturated groundwater, the MCL will not be exceeded for either stagnation or flowing water conditions for both SBR and NBR.

It is clear from these studies that certain elastomers exhibit permeation in gasoline or solvent laden soils, and care should be taken when selecting elastomers in distribution system locations of concern. AWWA Standards provide the following language in standards having plastic or elastomeric materials:

> *Permeation. The selection of materials is critical for potable water, wastewater and reclaimed water service and distribution piping systems in locations where there is likelihood the piping system will be exposed to significant concentrations of pollutants composed of low-molecular-weight petroleum products or organic solvents or their vapors. Documented research has shown that piping system materials, such as polyethylene and polyvinyl chloride and elastomers used in gaskets and packing glands, are subject to permeation by lower-molecular-weight organic solvents or petroleum products. If a potable water, wastewater or reclaimed water piping system must pass through such a contaminated area or an area subject to contamination, consult the manufacturer regarding permeation of pipe walls, valve components, jointing materials, and other piping system components, before selecting materials for use in that area.*

CHEMICAL RESISTANCE GUIDES

Many component manufacturers publish chemical resistance guides, which are helpful in predicting the resistance of elastomers to different conditions (Pruett 2014). Mixtures and other concentrations of chemicals are beyond the scope of most guides. Performances of materials in existing systems also offer valuable information in predicting product performance. The information given in Table 5-4 is a compilation of chemical resistance of elastomers at ambient temperature and consists of a letter designating the effectiveness of the elastomer as follows:

- A = resistant under ambient conditions
- B = conditional resistance, review performance
- C = not recommended
- AB = reported data varies widely
- X = no data available

These guides are based on scientific research and field experience and are a valuable tool in selecting elastomers for special service applications.

Table 5-4 Chemical resistance guide*

Chemical Resistance Guide
for **Generic Waterworks Elastomers**

A = resistant under ambient conditions, 70°F (21°C)
B = conditional resistance, review performance
C = not recommended
AB = reported data varies widely
X = no data available

Chemical	NBR – Acrylonitrile Butadiene (Nitrile, Buna-N)	CR – Chloroprene (Neoprene)	EPDM – Ethylene Propylene Diene Monomer	FKM – Fluorocarbon (Viton™, Teflon)	CSM – Chlorosulfonated Polyethylene (Hypalon™)	NR, IR – Polyisoprene (Natural Rubber)	SBR – Styrene Butadiene	VMQ – Polysiloxane (Silicone)	IIR – Isobutylene Isoprene (Butyl)	AU, EU – Polyurethane
Acetic acid (5%)	B	C	A	B	A	B	B	B	B	B
Acetone	C	C	A	C	C	B	C	C	AB	C
Alum (10%)	A	A	A	A	A	A	A	A	A	C
Aluminum chloride	A	A	A	A	A	A	A	A	A	B
Aluminum fluoride	A	A	A	A	A	B	A	B	A	C
Aluminum hydroxide	A	A	A	A	AB	A	B	A	A	A
Aluminum nitrate	A	A	A	A	A	A	A	B	A	C
Ammonia anhydrous	AB	A	A	C	B	C	B	C	A	C
Ammonium carbonate	AB	A	A	A	AB	A	A	BC	A	A
Ammonium chloride	AB	A	A	A	A	A	A	BC	A	AB
Ammonium hydroxide (30%)	A	A	A	AB	A	B	C	BC	A	C
Ammonium nitrate	A	BC	A	A	A	A	A	BC	A	B
Ammonium phosphate	A	A	A	A	A	A	A	A	A	B
Ammonium sulfate	A	A	A	A	A	A	A	A	A	C
Ammonium sulfite	A	A	A	A	A	A	A	A	A	C
Beer	A	AB	A	A	A	A	A	A	A	B
Benzene	C	C	C	B	C	C	C	C	AB	C
Bleach (5%)	C	C	A	A	A	C	B	B	AB	C
Calcium carbonate	A	A	A	A	A	A	A	B	A	C
Calcium chloride	A	A	A	A	A	A	A	A	A	C
Calcium hydroxide	A	A	A	A	A	A	A	A	A	C
Calcium hypochlorite	C	C	A	A	A	C	A	A	A	AB
Carbon dioxide (wet)	A	B	AB	AB	AB	AB	B	A	B	A
Carbon monoxide (gas)	A	A	A	A	A	C	BA	A	A	A
Caustic potash – KOH	A	AB	A	AB	A	B	B	C	A	C
Caustic soda – NaOH (20%)	A	A	A	B	A	A	B	A	A	AB
Chlorine (liq)	C	C	C	A	C	C	C	C	C	C
Copper sulfate (5%)	A	A	A	A	A	B	B	C	A	AB
Diesel fuel	A	C	C	A	C	C	C	C	C	C

*Adapted from Pruett (1994).

Chemical Resistance Guide
for **Generic Waterworks Elastomers**

A = resistant under ambient conditions, 70°F (21°C)
B = conditional resistance, review performance
C = not recommended
AB = reported data varies widely
X = no data available

Chemical	NBR – Acrylonitrile Butadiene (Nitrile, Buna-N)	CR – Chloroprene (Neoprene)	EPDM – Ethylene Propylene Diene Monomer	FKM – Fluorocarbon (Viton™, Teflon)	CSM – Chlorosulfonated Polyethylene (Hypalon™)	NR, IR – Polyisoprene (Natural Rubber)	SBR – Styrene Butadiene	VMQ – Polysiloxane (Silicone)	IIR – Isobutylene Isoprene (Butyl)	AU, EU – Polyurethane
Ferric chloride (sat)	A	A	A	A	A	A	A	A	A	A
Ferric hydroxide	A	AB	A	A	AB	C	X	X	A	X
Ferric sulfate	A	A	A	A	A	A	A	B	A	A
Ferrous sulfate	A	A	A	A	A	A	A	B	A	A
Fluorosilicic acid	A	A	A	A	A	A	AB	A	A	B
Formaldehyde (40%)	B	C	A	A	A	A	B	B	A	C
Gasoline	A	B	C	A	C	C	C	C	C	A
Glucose	A	A	A	A	A	A	A	A	A	A
Grease	A	B	C	A	C	C	C	C	C	A
Hydrazine	B	B	A	C	B	C	B	B	A	C
Hydrocyanic acid	B	B	A	A	A	AB	B	AC	AB	C
Hydrogen (gas)	A	A	A	A	A	A	B	A	A	A
Hydrogen peroxide (50%)	C	C	B	A	A	B	B	A	B	AB
Hydrogen sulfide (gas)	A	A	A	C	A	A	A	C	A	A
Hydrogen sulfide (wet)	B	A	A	C	A	C	C	A	A	A
Isobutyl alcohol	B	A	A	A	A	A	A	A	A	C
Isopropyl alcohol	A	AB	A	A	A	A	A	A	A	C
Ketones	C	C	A	C	B	A	AB	A	A	B
Magnesium hydroxide	A	A	A	A	A	A	B	AB	A	A
Manganese sulfate	A	A	A	A	A	AB	B	A	A	B
Methane	A	AB	C	A	A	C	B	A	C	C
Methyl ethyl ketone	C	C	A	C	C	C	A	B	A	C
Natural gas	A	A	C	A	AB	B	BC	A	C	BC
Nitric acid (10%)	C	C	B	A	AB	C	C	C	C	B
Nitrobenzene	A	C	C	A	C	C	C	C	B	C
Nitrogen (gas)	A	A	A	A	A	A	A	A	A	A
Nitrous oxide	A	B	A	A	A	A	A	X	A	AB
Oxygen (gas)	B	A	A	A	AB	B	B	A	A	A
Ozone (gas)	C	C	A	A	A	C	C	A	B	A

Chemical Resistance Guide
for **Generic Waterworks Elastomers**

A = resistant under ambient conditions, 70°F (21°C)
B = conditional resistance, review performance
C = not recommended
AB = reported data varies widely
X = no data available

Chemical	NBR – Acrylonitrile Butadiene (Nitrile, Buna-N)	CR – Chloroprene (Neoprene)	EPDM – Ethylene Propylene Diene Monomer	FKM – Fluorocarbon (Viton™, Teflon)	CSM – Chlorosulfonated Polyethylene (Hypalon™)	NR, IR – Polyisoprene (Natural Rubber)	SBR – Styrene Butadiene	VMQ – Polysiloxane (Silicone)	IIR – Isobutylene Isoprene (Butyl)	AU, EU – Polyurethane
Potassium nitrate	A	A	A	A	A	A	A	A	A	A
Potassium permanganate	C	A	A	A	AB	A	AB	X	A	B
Propane	A	C	C	A	B	C	C	C	C	B
Sodium aluminate	A	A	A	A	A	AB	AB	X	A	X
Sodium bicarbonate	A	A	A	A	A	A	A	A	A	A
Sodium carbonate	A	A	A	A	A	A	A	A	AB	A
Sodium chloride	A	A	A	A	A	A	A	A	A	A
Sodium fluoride	A	A	A	A	AB	B	AB	X	A	AB
Sodium hydroxide (15%)	A	A	A	B	A	A	B	A	A	C
Sodium hypochlorite	C	C	A	A	A	C	B	B	B	B
Sodium metaphosphate	A	B	A	A	B	A	A	A	AB	B
Sodium perborate	B	B	A	A	B	B	B	B	A	B
Sodium peroxide	B	B	A	A	B	B	B	C	A	C
Sodium silicate	A	A	A	A	A	A	A	A	A	AB
Sulfur dioxide (dry)	C	C	A	A	A	B	BC	AB	B	B
Toluene	C	C	C	A	C	C	C	C	C	C
UV radiation	C	B	B	A	A	C	A	A	A	A
Vegetable oil	A	B	B	A	B	C	C	A	AC	A
Water, brackish	A	A	A	A	A	A	A	A	A	A
Water, deionized	A	B	A	A	A	A	X	C	A	A
Water, potable	A	A	A	A	A	A	A	A	A	A
Water, chloramines 1% (Rockaway et al. 2007)	B	C	A	A	B	C	B	A	A	X
Water, chlorinated 1% (Rockaway et al. 2007)	A	B	A	A	B	B	B	A	A	X
Water, sea	A	A	A	A	A	A	A	A	A	A
Water, waste	A	A	C	B	A	AB	AB	AB	AB	C
Xylene	C	C	C	A	C	C	C	C	C	C
Zinc chloride	A	A	A	A	A	A	A	AB	A	AB

CONCLUSION

Understanding and predicting elastomer degradation is important to the water industry. Elastomers that degrade more quickly than expected create service problems, making it difficult for utilities and infrastructure owners to cost and effectively plan preventive maintenance programs, and can negatively affect customer relations. Because water distribution systems are basically sealed, the speed and consequences of degradation cannot be monitored. Moreover, because water distribution systems are unique in design and treatment application, generalizations relative to elastomer performance are difficult to make. Therefore, material choices and failure modes of components must be evaluated based on field experience and research-based expectations.

REFERENCES

Bonds, R.W. 2004. Effect of Chloramines on Ductile-Iron Pipe Gaskets of Various Elastomer Compounds. *Journal AWWA*, 96:4:153.

Cheng, C.L.; Gaunt, J.A.; Mao, F.; and Ong, S.K. 2012. Permeation of Gasoline Through DI Pipe Gaskets in Water Mains. *Journal AWWA*.

Glaza, E.C. and Park, J.K. 1992. Permeation of Organic Contaminants Through Gasketed Pipe Joints. *Journal AWWA*, 92:7:92.

Kirmeyer, G.J.; Foust, G.W.; Pierson, G.L.; Simmler, J.J.; and LeChevevallier, M.W. 2004 (2nd ed.). Optimizing Chloramine Treatment. Denver, Colo.: AWWA Research Foundation.

Mao, F.; Gaunt, J.A.; Ong, S.K.; and Cheng, C.L. 2011. Permeation of Petroleum-Based Hydrocarbons through PVC Pipe Joints with Rieber Gasket Systems. *Journal of Environmental Engineering*, December.

Ong, K.; Gaunt, J.A.; Mao, F.; Cheng, C.-L.; Esteve-Agelet, L.; and Hurburgh, C.R. 2008. Impact of Hydrocarbons on PE/PVC Pipes and Pipe Gaskets. Denver, Colo.: AWWA Research Foundation.

Pruett, K. 2014. *Chemical Resistance Guide for Elastomers IV.* California: Compass Publications.

Rahman, S. 2007. Sealing Our Buried Lifelines. *AWWA OPFLOW*, April.

Reiber, S. 1993. Investigating the Effects of Chloramine on Elastomer Degradation. *Journal AWWA*, 85:8:101–111.

Rockaway, T.D.; Willing, G.A.; Schreck, R.M.; and Davis, K.R. 2007. *Performance of Elastomeric Components in Contact with Potable Water*. Denver, Colo.: AWWA Research Foundation.

San Francisco Public Utilities Commission. 2006. sfwater.org/detail.cfm/MC_ID/13/MSC_ID/166/MTO_ID/399/C_ID/2215> (cited April 29, 2006).

Sarai, D.S. 2005. Basic Chemistry for Water and Wastewater Operators. Denver, Colo.: AWWA.

US Environmental Protection Agency (USEPA). Chloramines in Drinking Water. Washington DC: USEPA.

This page intentionally blank.

Chapter 6

Storage and Maintenance of Elastomers

Elastomeric parts or products containing elastomeric components must be maintained in stock by water distribution system owners. When a field emergency such as a line break occurs, the team of responders will need replacement valves, fittings, and gaskets to make the necessary repairs. Distribution systems also have complex equipment, such as pumping systems, that require regular maintenance and repair. To efficiently accomplish these repairs, a store room of spare seals, pipe gaskets, and valve parts are typically kept on hand at all times. However, because elastomers are subject to aging and degradation over time, storage of parts should be controlled with respect to time and the environment in which the parts reside.

Physical changes to elastomers can occur due to ultraviolet (UV) light, temperature exposure, ozone exposure, and aging but are difficult to predict. However, with certain care, the life of replacement elastomeric parts can be maximized. The replacement intervals of elastomeric parts in service should be recorded as a predictor to prevent future failures.

STORAGE AND HANDLING

Elastomers in products may undergo changes in their physical properties. Over time, they can become unusable due to excessive hardening, softening, cracking, crazing, or other surface degradations. These changes may be the result of a single factor or a combination of factors, such as the action of oxygen, ozone, sunlight, heat, humidity, oils, water, or other solvents. These detrimental effects can be minimized by proper storage conditions.

New products should be stored indoors or covered before installation, if possible. Exposing elastomeric seals to outdoor conditions with UV sunlight or ozone and elevated temperatures greater than 100°F (38°C) can rapidly degrade the elastomeric parts. The ideal storage temperature range is between 40°F (4°C) and 80°F (26°C). Lower or freezing temperatures are typically not deleterious to the elastomers. Similarly, humid environments are acceptable.

If the products containing elastomeric seals must be stored outside, they should be secured with flange covers and dark-colored shrink wrap to minimize exposure to the elements and keep the parts free from debris. Elastomeric parts should be stored in the relaxed condition. Therefore, depending on the design, products containing elastomeric seals are typically shipped in the partially open position. Parts should not be stored near ozone generators, including electric motors, mercury vapor lamps, and high-voltage equipment.

For long-term storage (i.e., greater than six months), consult the recommendations of the manufacturer. Should the manufacturer not have long-term storage recommendations, store products in a warehouse with a controlled environment, if possible. High temperatures can cause premature aging of valve seats and increase their hardness. Exposure to ozone from the atmosphere or nearby electrical equipment can cause surface cracking. To prevent surface degradation, protective spray coatings are available for rubber surfaces, but they should be approved by the manufacturer prior to use and be compatible with the elastomer. When properly stored, elastomeric parts should perform as new for up to 5–10 years, depending on the elastomer.

A recordkeeping and tagging system is important to control the inventory of replacement parts. The oldest parts should be used first, and if the age is greater than 5 years, an inspection of the part should be undertaken.

When installing fittings and valves, forks, chains, or slings should not be put through the port opening or elastomeric surfaces or mating surfaces. They can be easily damaged during lifting. Try to move valves and fittings using rods or eye bolts inserted in the flange holes or lifting eyes provided on the product.

LUBRICATION AND MAINTENANCE

Most elastomeric parts, such as O-ring seals and shaft packing, are factory-lubricated to reduce wear and friction in service. Certain pipe joints, such as push-on and mechanical joints, require a lubricant to be applied to the gaskets. The suggested lubricant from the manufacturer should be used because some lubricants can be harmful to certain elastomers. The lubricant must be nontoxic, not support the growth of bacteria, have no deteriorating effects on the gasket material, not impart taste or odor to the water in the pipe, and be safe for potable water consumption.

INSPECTION

The condition and performance of an elastic part can be predicted by a close examination of the part. When degradation occurs, it can usually be observed on the surface. For example, the surface irregularities or checking shown in Figure 6-1 are an indication of exposure to UV light or ozone and could significantly affect the performance of this elastomeric part in service.

If the original properties of the part are known, the hardness can be checked with an economical pocket durometer gauge (Figure 6-2), and compared with the original hardness value. An increase in hardness is an indication of aging and loss of elasticity.

Figure 6-1 Surface degradation of an elastomeric part

Figure 6-2 Portable rubber hardness tester

The manufacturer of the elastomeric part can provide additional insight into the evaluation of aged parts.

This page intentionally blank.

Appendix **A**

Table of Physical/ Mechanical Properties of Gaskets for Various Pipe Material Standards

Table A-1 Physical/mechanical properties of gaskets for plastic, ductile iron, steel, and clay pipe

Physical/Mechanical Property	Plastic Pipe		Ductile Iron Pipe	Steel Pipe	Clay Pipe
	ASTM F477-14 Gaskets for Joining Plastic Pipes (Thermoplastics and Thermosets, e.g., PVC, Corrugated HDPE, Fiberglass Pipes)	ASTM F477-14 (Oil-Resistant) Gaskets for Joining Plastic Pipes (Thermoplastics and Thermosets, e.g., PVC, Corrugated HDPE, Fiberglass Pipes)	AWWA C111-17 Gasket Joints for Ductile Iron Pipe and Fittings	AWWA C200-17 Steel Water Pipe	ASTM C425-04 (Reapproved 2018) Standard Specification for Compression Joints for Vitrified Clay Pipe and Fittings Compression Joints for Vitrified Clay Pipe
Elastomer formulation	The basic polymer shall be natural rubber, synthetic rubber, a blend of both, or a thermoplastic elastomer.	The compound shall contain not less than 50% by volume of oil-resistant polymer.	Rubber gaskets shall be made of vulcanized SBR unless otherwise specified by the purchaser, such as Ethylene Propylene Diene Monomer (EPDM), nitrile, and fluorocarbons. Reclaimed or natural rubber shall not be used.	The rubber compound shall contain not less than 50% by volume of first-grade synthetic rubber or synthetic rubber blends. The remainder of the compound shall consist of pulverized fillers free from rubber substitutes, reclaimed rubber, and deleterious substances.	No guidance on elastomer formulation provided in Standard.
Tensile strength	1,200 psi for low-head (LH) gaskets 2,000 psi for high-head (HH) gaskets / 1,500 psi for EPDM and nitrile HH gaskets	1,200 psi for LH gasket 1,500 psi for HH gasket	2,000 psi min	2,300 psi min	1,500 psi min
Elongation @ break	325% for LH 400% for HH	325% for LH 350% for HH	300% min	350% min	500% min
Compression set	25% LH / 20% HH	25% LH / 20% HH	20% max	20% max	16% max
Hardness type A durometer	40–60	40–60	50–65 ± 5	50–65 ± 5	35–50
Ozone resistance	No cracks 50 pphm / 70 h	No cracks 50 pphm / 70 h	No cracks 50 pphm / 25 h	No cracks 50 pphm / 72 h	No cracks 50 pphm / 24 h
Volume change – Water immersion	5% max	10% max	–	5% max	5% max
Weight change – Water immersion	–	–	–	–	–
Accelerated – aging tensile retained	85% min	85% min	60% min	85% min	80% min
Accelerated – aging elongation retained	80% min	80% min	60% min	80% min	75% min
Hardness change after aging	+8 pts	+8 pts	–	–	–
Volume change – Oil immersion	–	50% max	–	–	–
Density	–	–	–	0.95–1.45 ± 0.05	–
Splice strength classification	–	–	–	–	–

Adapted from ASTM International.

Table A-2 Physical/mechanical properties of gaskets for concrete pressure pipe

	Concrete Pressure Pipe		
Physical/ Mechanical Property	AWWA C301-14 Prestressed Concrete Cylinder Pipe (PCCP)	AWWA C302-16 Reinforced Concrete Non-Cylinder Pipe	AWWA C303-17 Bar-Wrapped, Steel Cylinder Pipe
Elastomer formulation	The rubber compound shall contain not less than 50% by volume of polyisoprene or other synthetic rubbers.	The rubber compound shall contain not less than 50% by volume of polyisoprene or synthetic rubbers. The remainder of the compound shall consist of pulverized fillers free from rubber substitutes, reclaimed rubber, and deleterious substances.	The rubber compound shall contain not less than 50% by volume of polyisoprene or other synthetic rubber. The remainder of the compound shall consist of pulverized fillers free from rubber substitutes, reclaimed rubber, and deleterious substances.
Tensile strength	2,700 psi min (compound) – polyisoprene gaskets	2,700 psi min (compound) – polyisoprene gaskets	2,700 psi min (compound) – polyisoprene gaskets
	2,000 psi min (compound) – other synthetic rubber gaskets	2,000 psi min (compound) – synthetic rubber gaskets	2,000 psi min (compound) – other synthetic rubber gaskets
Elongation @ break	400% min – polyisoprene gaskets	400% min – polyisoprene gaskets	400% min – polyisoprene gaskets
	350% min – other synthetic rubber gaskets	350% min – synthetic rubber gaskets	350% min – other synthetic rubber gaskets
Compression set	20% max	20% max	20% max
Hardness type A durometer	50–65	30–50 (concrete spigot) 50–65 (steel spigot)	50–65
Ozone resistance	–	–	No cracks
	–	–	50 pphm / 72 h
Volume change – Water immersion	–	–	5% max
Weight change – Water immersion	–	–	–
Accelerated – aging tensile retained	80% min (compound)	80% min (compound)	80% min (compound)
Accelerated – aging elongation retained	–	–	–
Hardness change after aging	–	–	–
Volume change – Oil immersion	–	–	–
Density	0.95–1.45 ± 0.05	0.95–1.45 ± 0.05	0.95–1.45 ± 0.05
Splice strength classification	–	–	–

Adapted from ASTM International.

Table A-3 Physical/mechanical properties of gaskets for concrete gravity pipe

Physical/Mechanical Property	Concrete Gravity Pipe			
	ASTM C443-12 (Reapproved 2017) Joints for Concrete Pipe and Manholes, Using Rubber Gaskets	ASTM C443-12 (Reapproved 2017) (Oil-Resistant) Joints for Concrete Pipe and Manholes, Using Rubber Gaskets	ASTM C361 Reinforced Concrete Low-Head Pressure Pipe	ASTM C361 (Oil-Resistant) Reinforced Concrete Low-Head Pressure Pipe
Elastomer formulation	The basic polymer shall be natural rubber, synthetic rubber, or a blend of both.	The compound shall contain not less than 50% by volume of oil-resistant polymer	The basic polymer shall be natural rubber, synthetic rubber, or a blend of both.	The compound shall contain not less than 50% by volume of oil-resistant polymer
Tensile strength	1,200 psi min	1,200 psi min	2,300 psi min	1,500 psi min
Elongation @ break	350% min	350% min	425% min	350% min
Compression set	25% max	25% max	20% max	20% max
Hardness type A durometer	40–60	40–60	35–50 ± 5 (concrete spigot) 35–65 ± 5 (steel spigot)	40–65 ± 5
Ozone resistance	No cracks	No cracks	No cracks	No cracks
	50 pphm / 72 h	50 pphm / 72 h	50 pphm / 72 h	50 pphm / 72 h
Volume change – Water immersion	–	–	–	–
Weight change – Water immersion	10% max	15% max	5% max	15% max
Accelerated – aging tensile retained	85% min	80% min	85% min	80% min
Accelerated – aging elongation retained	80% min	60% min	80% min	80% min
Hardness change after aging	–	+15 pts	–	+15 pts
Volume change – Oil immersion	–	80% max	–	80% max
Density	–	–	–	–
Splice strength classification	Class 3	Class 2	Class 3	Class 2

Adapted from ASTM International.

REFERENCES

American Water Works Association (AWWA). 2017. AWWA C111/A21.11, Rubber-Gasket Joints for Ductile-Iron Pressure Pipe and Fittings. Denver, Colo.: AWWA.

AWWA. 2017. AWWA C200. Steel Water Pipe – 6 in. (150 mm) and Larger. Denver, Colo.: AWWA.

AWWA. 2014. AWWA C301, Prestressed Concrete Pressure Pipe, Steel-Cylinder Type. Denver, Colo.: AWWA.

AWWA. 2016. AWWA C302, Reinforced Concrete Pressure Pipe, Noncylinder Type. Denver, Colo.: AWWA.

AWWA. 2017. AWWA C303, Concrete Pressure Pipe, Bar-Wrapped, Steel-Cylinder Type. Denver, Colo.: AWWA.

American Society for Testing and Materials (ASTM) International. 2016. ASTM C361, Standard Specification for Reinforced Concrete Low-Head Pressure Pipe, Philadelphia, Pa.: ASTM International.

ASTM International. 2004. ASTM C424, Standard Specification for Compression Joints for Vitrified Clay Pipe and Fittings, Philadelphia, Pa.: ASTM International.

ASTM International. 2017. ASTM C443, Standard Specification for Joints for Concrete Pipe and Manholes, Using Rubber Gaskets, Philadelphia, Pa.: ASTM International.

ASTM International. 2014. ASTM F477, Standard Specification for Elastomeric Seals (Gaskets) for Joining Plastic Pipe, Philadelphia, Pa.: ASTM International.

This page intentionally blank.

Appendix **B**

Elastomer Test Protocols

PERCENT ASH, PERCENT CARBON BLACK (ASTM D297)

Standard Test Methods for Rubber Products—Chemical Analysis

These test procedures provide quantitative and qualitative analysis (including ash analysis and fillers analysis, including percent carbon black) of the composition of rubber products. Many of these test routines can be used for the analysis of synthetic and natural crude rubbers.

COMPRESSION SET METHOD B (ASTM D395)

Standard Test Methods for Rubber Property—Compression Set

These test procedures outline the testing of rubber to be used in applications where the rubber will be subjected to compressive stresses in liquid media or air. The test method applies to the rubber used in seals, vibration dampers, and machinery mountings. Method B is a test involving compression set under constant force in air.

TENSILE, ELONGATION, AND MODULUS PROPERTIES (ASTM D412)

Standard Test Methods for Vulcanized Rubber and Thermoplastic Elastomers—Tension

These test procedures deal with methods used to evaluate the tensile (tension) properties of thermoplastic elastomers and vulcanized thermoset rubbers. There are two procedures—Test Procedure A covers dumbbell and straight section samples, while Test Procedure B covers cut ring samples.

ADHESION TO A FLEXIBLE SUBSTRATE (ASTM D413)

Standard Test Methods for Rubber Property—Adhesion to a Flexible Substrate

These test procedures deal with the determination of the adhesion strength between plies of fabric bonded with rubber or the adhesion of the rubber layer in items made from rubber then attached to an other material. These test methods apply when the adhered surfaces are uniformly circular or approximately planar as in hose, belting, rubber-covered sheet metal, or tire carcasses.

ADHESION TO A RIGID SUBSTRATE (ASTM D429)

Standard Test Methods for Rubber Property—Adhesion to Rigid Substrates

These test procedures deal with routines for testing the static adhesion strength of rubber to rigid materials (in most cases, metals). Procedure A is the method for a rubber part assembled in between two parallel metal plates. Procedure B is the routine for perpendicular stripping of a test-rubber part assembled to a single metal plate. Procedure C is used to measure the adhesion of rubber to metal with a conical sample. Procedure D involves the adhesion test post-vulcanization (PV) bonding of rubber to metal. Procedure E is the routine for perpendicular stripping of a test rubber tank lining assembled to a single metal plate. Procedure F involves a rubber part assembled in between two parallel convex-shaped metal plates. Procedure G is used to measure bond durability for rubber-to-metal bonded components with a double shear cylindrical sample. Procedure H is used in measuring bond durability for rubber-to-metal bonded parts with a quadruple shear cylinder sample.

CHEMICAL RESISTANCE (CHANGE IN HARDNESS AND VOLUME/70 H) AND OIL RESISTANCE (70 H) (ASTM D471)

Standard Test Method for Rubber Property—Effect of Liquids

This test procedure deals with the required methods to evaluate the comparative ability of rubber-like and rubber compositions to withstand the effect of liquids. It is designed for testing samples cut from fabric coated with vulcanized rubber, samples of vulcanized rubber cut from standard sheets, or finished articles of commerce.

DETERIORATION IN AN AIR OVEN (ASTM D573)

Standard Test Method for Rubber—Deterioration in an Air Oven

This test procedure outlines a method to determine the influence of elevated temperature on the physical properties of vulcanized rubber. This test routine can be used to evaluate rubber compounds on a laboratory comparison basis.

COMPRESSION DEFLECTION (ASTM D575)

Standard Test Methods for Rubber Properties in Compression

These test procedures outline the test methods for determining the compression-deflection characteristics of rubber compounds other than those usually classified as sponge rubber and hard rubber.

TEAR RESISTANCE (ASTM D624)

Standard Test Method for Tear Strength of Conventional Vulcanized Rubber and Thermoplastic Elastomers

This test procedure outlines the methods for measuring a property of thermoplastic elastomers and conventional vulcanized thermoset rubber referred to as tear strength.

CHANGE IN HARDNESS (ASTM D865)

Standard Test Method for Rubber Deterioration by Heating in Air (Test Tube Enclosure)

This test procedure outlines a method to determine the deterioration induced by heating rubber samples in individual test tube enclosures with circulating air. This isolation prevents cross-contamination of compounds due to loss of volatile materials (for example, antioxidants) and their subsequent migration into other rubber compounds. The absorption of such volatile materials can influence the rate of degradation of rubber compounds.

HARDNESS (DUROMETER) (ASTM D2240)

Standard Test Method for Rubber Property—Durometer Hardness

This test procedure outlines 12 types of rubber hardness measurement devices known as durometers: Types A and D. The procedure for determining indentation hardness of substances classified as vulcanized (thermoset) rubber, thermoplastic elastomers, cellular materials, elastomeric materials, gel-like materials, and some plastics is also described.

IDENTIFICATION BY INFRARED (ASTM D3677)

Standard Test Methods for Rubber—Identification by Infrared Spectrophotometry

These test procedures are based on the infrared examination of films, and pyrolysis products (pyrolyzates) can be used for rubber identification. Additionally, these test routines apply to rubbers in the raw state and, when compounded, both in the uncured and cured states.

EFFECTS OF CHLORINE AND CHLORAMINE AQUEOUS SOLUTIONS (ASTM D6284)

Standard Test Methods for Rubber Property—Effect of Aqueous Solutions with Available Chlorine and Chloramine

This test method covers procedures for evaluating the ability of rubber and rubber-like materials to withstand the effects of aqueous solutions with available chlorine and chloramine. It is intended to compare the effects of chlorine compounds, present in potable water due to disinfection procedures, on rubber articles. Test solutions are designed to contain chlorine compounds, including hypochlorous acid (HOCl), hypochlorite ions (OCl), and monochloramine (NH_2Cl). The standard does not provide pass/fail criteria and should only be used to compare different elastomers.

REFERENCES

American Society for Testing and Materials (ASTM) International. 2015. ASTM D297, Standard Test Methods for Rubber Products—Chemical Analysis. Philadelphia, Pa.: ASTM International.

ASTM International. 2016. ASTM D395, Standard Test Methods for Rubber Property—Compression Set. Philadelphia, Pa.: ASTM International.

ASTM International. 2016. ASTM D412, Standard Test Methods for Vulcanized Rubber and Thermoplastic Elastomers—Tension. Philadelphia, Pa.: ASTM International.

ASTM International. 2017. ASTM D413-98, Standard Test Methods for Rubber Property—Adhesion to Flexible Substrate. Philadelphia, Pa.: ASTM International.

ASTM International. 2014. ASTM D429, Standard Test Methods for Rubber Property—Adhesion to Rigid Substrates. Philadelphia, Pa.: ASTM International.

ASTM International. 2016. ASTM D471, Standard Test Methods for Rubber Property—Effects of Liquids. Philadelphia, Pa.: ASTM International.

ASTM International. 2015. ASTM D573-04, Standard Test Methods for Rubber—Deterioration in an Air Oven. Philadelphia, Pa.: ASTM International.

ASTM International. 2012. ASTM D575, Standard Test Methods for Rubber Properties in Compression. Philadelphia, Pa.: ASTM International.

ASTM International. 2012. ASTM D624, Standard Test Methods for Tear Strength of Conventional Vulcanized Rubber and Thermoplastic Elastomers. Philadelphia, Pa.: ASTM International.

ASTM International. 2015. ASTM D2240, Standard Test Methods for Rubber Property—Durometer Hardness. Philadelphia, Pa.: ASTM International.

ASTM International. 2015. ASTM D3677, Standard Test Methods for Rubber—Identification by Infrared Spectrophotometry. Philadelphia, Pa.: ASTM International.

ASTM International. 2017. ASTM D6284, Standard Test Methods for Rubber Property—Effect of Aqueous Solutions with Available Chlorine and Chloramine. Philadelphia, Pa.: ASTM International.

Index

Note: *f.* indicates figure; *t.* indicates table

A

accelerants, 35
accelerated-aging test protocols, 52
acrylonitrile butadiene (NBR; Nitrile, Buna-N), 21, 23, 55
activators, 35
additives, 2, 35
adhesion, 4
Adhesion to a flexible substrate (ASTM D413), 72
Adhesion to a rigid substrate (ASTM D429), 72
Aerospace Standard AS-568, 9
age resistance, 4
antidegradants, 4, 35
antioxidants, 35
Arrhenius equation, 53
ASTM D2000, 22–23
  definitions of material types and classes, 23*t.*
ASTM F477, 12
autoclave, 4, 26
AWWA C207, 11–12

B

benzene, toluene, ethylbenzene, and xylene (BTEX), 54
bond disassociation energy, 48, 49*t.*

C

calcium oxide, 35
carbon black, 4, 35
Change in hardness (ASTM D865), 73
checking, 4
chemical resistance, 4
  ASTM D471, 72
  guides, 55–58, 56*t.*–58*t.*
  of elastomers, 47–59
chloramination, 51
chloramine, 50, 53
  susceptibility of elastomers without antidegradants chloramine attach, 52*t.*
Chloramine Effects on Distribution System Materials, 51–52
chlorine, 50, 53
chloroprene rubber (CR; Neoprene), 20, 47
chlorosulfonated polyethylene (CSM; Hypalon®), 20
compound, 4, 33
Compression deflection (ASTM D575), 72
compression molding, 27
compression set, 4, 40–41, 44
  apparatus, 41*f.*
Compression set method B (ASTM D395), 71
copolymer, 4
cracking, 4
crazing, 4
cross-linkages, 47, 48

D

degradation, 1
  definition of, 4
  trends, of elastomers, 51–54
    damage to an encapsulated gate valve wedge, 51*f.*
    generic elastomers without antidegradants in chloramine service, 53*t.*
desiccants, 35
Deterioration in an air oven (ASTM D573), 72
diaphragms, 17–18
  globe control valve, 17*f.*
dichloramine, 50
disinfectant, 4
  chlorine and chloramine, 50
drinking water system, 36, 54
  compliance, 49–50

ductile iron pipe, 12, 12*f.*, 54
durometer, 4, 37–38
  device, 38*f.*
  of elastomers at reduced temperatures, 43*f.*
  various scales, 38*f.*
dynamic seals, 14–17
  epoxy-restrained resilient seat in AWWA butterfly valve, 15*f.*
  in dry-barrel fire hydrant, 15*f.*
  knife gate valve, 16, 16*f.*
  typical bolted resilient seat in AWWA butterfly valve, 15*f.*

E

Effects of chlorine and chloramine solutions (ASTM D6284), 73
elastomer test protocols, 45*t.*, 71–73
elastomer test specimen, 38, 39*f.*
elastomeric seals, 3*f.*, 14, 62
elastomers
  chemical resistance, 47–59
  chemistry of, 47–49
  classifications, 2
  common ingredients and their functions, 35*t.*
  definition of, 4
  degradation
    of a check valve disc in potable water service, 48*f.*
    trends, 51–54, 51*f.*
  history of, 2–3
  manufacture of seals and gaskets, 25–30
  mechanical properties, 33–45
  permeation of, 54–55
  selection of, 24–25
    common elastomers and generic properties, 24*t.*
  storage and maintenance, 61–63
    inspection, 62–63
    lubrication and maintenance, 62
    storage and handling, 61–62
  types of, 19–23
  typical operating temperature range, 43*f.*
  use of, 7–18
    diaphragms, 17–18, 17*f.*
    dynamic seals, 14–17
    linings and coatings, 7–8
    static seals, 8–14
elongation, 40
ethylene propylene diene monomer (EPDM), 20, 36, 42, 48–49, 54

F

fillers, 35
flange gaskets, 9–12, 54
  sample full-face and ring-type gaskets, 11*f.*
  typical waterworks flange, 10*f.*
  waterworks check valve flange connection with full-face gasket, 10*f.*
  waterworks flange and fittings standards, 10*t.*
flashing, 4, 27
fluorocarbon (FKM; Viton®), 20–21, 52
formulation, 4
  example for SBR compound, 22*t.*
full-face gasket, 11, 11*f.*

G

gaskets, 10
  manufacture of, 25–30
  physical/mechanical properties
    for concrete gravity pipe, 68*t.*
    for concrete pressure pipe, 67*t.*
    for plastic, ductile iron, steel, and clay pipe, 66*t.*

H

hardness, 4, 11, 37
  ASTM D2240, 73
Identification by infrared spectrophotometry (ASTM D3677), 73
injection molding, 27
  fabrication of seals using an extrusion die, 29*f.*
  typical process, 28*f.*
inspection, 62–63
isobutene isoprene (IIR; Butyl), 21

L

line call-out, 23
linings and coatings, 7–8
  check valve with rubber-lined interior and flange facing, 8*f.*
  gate valve wedge, 8*f.*
lubrication, 62

M

maximum contaminant level (MCL), 55

maximum residual disinfectant level (MRDL), 49
mechanical joint gaskets, 14, 14*f.*, 54
mill, 4, 25
monochloramine, 50

N

nitrile (NBR), 34, 42, 47, 54
  composition of, 34*f.*
  typical heat aging test variation, 44*t.*
NPS 24 (DN 600) gray iron butterfly valve body, 26*f.*
NSF/ANSI/CAN 61, 36
  sample online certification, 37*f.*

O

O-rings, 8–9
  wedge gate valve with rubber encapsulated wedge, 8*f.*
    installation without and with pressure, 9*f.*
oakum, 4

P

Percent ash, percent carbon black (ASTM D297), 71
permeation, 4, 55
  of elastomers, 54–55
peroxide, 49
pigments, 35
pinch valve, 16, 17*f.*
pipe connections, 12–13
plastic, 4
plasticizers, 35
polyester, 20
polyether urethane, 20
polyethylene (PE) pipe, 12
  butt fusion of, 13*f.*
  electrofusion coupling for, 13*f.*
  mechanical joint adapter for, 13*f.*
polyisoprene (NR, IR; natural rubber, synthetic rubber), 21
polymer, 4
polysiloxane (VMQ; Silicone), 22
polyvinyl chloride (PVC) pipe, 12, 12*f.*
push-on gaskets, 12–13, 54
  PVC pipe push-on joint, 12*f.*
  required properties, 37*t.*

R

resilience, 4, 40
rheometer, 30, 30*f.*
ring gasket, 11, 11*f.*
rubber, 4
  lining of interior of a butterfly valve body, 26*f.*
  loading of ingredients, 26*f.*
  manufacturing considerations, 30–31
    rheometer fixture, 30*f.*
  natural, single monomer of, 34*f.*
  portable hardness tester, 63*f.*
  samples of molded parts, 29*f.*
  typical molding press with electric heating, 27*f.*
  vulcanization with sulfur, 34*f.*
rubber dumbbells, 38

S

safety, of elastomers, 35–36
seals, manufacture of, 25–30
seating stress, 11
Shore 00 durometers, 37
Shore A durometers, 37, 38*f.*
Shore D durometers, 37
static seals, 8–14
  flange gaskets, 9–12, 10*f.*, 10*t.*
  mechanical joint gaskets, 14
  O-rings, 8–9, 8*f.*, 9*f.*
  push-on gaskets and pipe connections, 12–13
  valve face seal, 14, 15*f.*
steel pipe, 12
  flanged and grooved joints, 13*f.*
strength, of elastomers, 36–40
styrene butadiene (SBR), 21 22, 47, 55
  sample formulation, 22*t.*
sulfur curing, 48
surface degradation, of an elastomeric part, 63*f.*

T

tear resistance, 41, 42*f.*
  ASTM D624, 73
tear strength, 5
temperature resistance, 42–44
  durometer of elastomers at reduced temperatures, 43*f.*

heat aging test variation for NBR, 44*t*.
operating temperature range of elastomers, 43*f*.
tensile strength, 5, 38
Tensile, elongation and modulus properties (ASTM D412), 71
tension testing machine, 38, 39*f*.
test protocols, 44–45
standard elastomer tests, 45*t*.
transfer molding, 27, 28*f*.
trichloramine, 50
trihalomethanes (THMs), 49

U

ultimate elongation, 5

V

valve face seal, 14
air/vacuum valve, 15*f*.
valve packing, 16
typical V-packing set in an AWWA butterfly valve, 17*f*.
vulcanization, 5

W

water sealing, 3

AWWA Manuals

M1, *Principles of Water Rates, Fees, and Charges*, #30001
M2, *Instrumentation and Control*, #30002
M3, *Safety Management for Water Utilities*, #30003
M4, *Water Fluoridation Principles and Practices*, #30004
M5, *Water Utility Management*, #30005
M6, *Water Meters—Selection, Installation, Testing, and Maintenance*, #30006
M7, *Problem Organisms in Water: Identification and Treatment*, #30007
M9, *Concrete Pressure Pipe*, #30009
M11, *Steel Pipe—A Guide for Design and Installation*, #30011
M12, *Simplified Procedures for Water Examination*, #30012
M14, *Backflow Prevention and Cross-Connection Control: Recommended Practices*, #30014
M17, *Fire Hydrants: Installation, Field Testing, and Maintenance*, #30017
M19, *Emergency Planning for Water and Wastewater Utilities*, #30019
M20, *Water Chlorination/Chloramination Practices and Principles*, #30020
M21, *Groundwater*, #30021
M22, *Sizing Water Service Lines and Meters*, #30022
M23, *PVC Pipe—Design and Installation*, #30023
M24, *Planning for the Distribution of Reclaimed Water*, #30024
M25, *Flexible-Membrane Covers and Linings for Potable-Water Reservoirs*, #30025
M27, *External Corrosion Control for Infrastructure Sustainability*, #30027
M28, *Rehabilitation of Water Mains*, #30028
M29, *Water Utility Capital Financing*, #30029
M30, *Precoat Filtration*, #30030
M31, *Distribution System Requirements for Fire Protection*, #30031
M32, *Computer Modeling of Water Distribution Systems*, #30032
M33, *Flowmeters in Water Supply*, #30033
M36, *Water Audits and Loss Control Programs*, #30036
M37, *Operational Control of Coagulation and Filtration Processes*, #30037
M38, *Electrodialysis and Electrodialysis Reversal*, #30038
M41, *Ductile-Iron Pipe and Fittings*, #30041
M42, *Steel Water-Storage Tanks*, #30042
M44, *Distribution Valves: Selection, Installation, Field Testing, and Maintenance*, #30044
M45, *Fiberglass Pipe Design*, #30045
M46, *Reverse Osmosis and Nanofiltration*, #30046
M47, *Capital Project Delivery*, #30047
M48, *Waterborne Pathogens*, #30048
M49, *Quarter-Turn Valves: Head Loss, Torque, and Cavitation Analysis*, #30049
M50, *Water Resources Planning*, #30050
M51, *Air Valves: Air-Release, Air/Vacuum and Combination*, #30051
M52, *Water Conservation Programs—A Planning Manual*, #30052
M53, *Microfiltration and Ultrafiltration Membranes for Drinking Water*, #30053
M54, *Developing Rates for Small Systems*, #30054
M55, *PE Pipe—Design and Installation*, #30055
M56, *Nitrification Prevention and Control in Drinking Water*, #30056
M57, *Algae: Source to Treatment*, #30057
M58, *Internal Corrosion Control in Water Distribution Systems*, #30058
M60, *Drought Preparedness and Response*, #30060
M61, *Desalination of Seawater*, #30061
M62, *Membrane Applications for Water Reuse*, #30062
M63, *Aquifer Storage and Recovery*, #30063

M65, *On-Site Generation of Hypochlorite,* #30065

M66, *Cylinder and Vane Actuators and Controls—Design and Installation,* #30066

M68, *Water Quality in Distribution Systems,* #30068

M69, *Inland Desalination and Concentrate Management,* #30069

M77, *Condition Assessment of Water Mains,* #30077